ATOMS, STARS AND MINDS

SYNTHESIZING AN ELEMENTARY PARTICLE THAT COMPREHENDS ITSELF

Peter Bros

1992

PUBLISHED BY FINANCIAL BOOK PARTNERS

PRINTED IN THE UNITED STATES OF AMERICA

ISBN 0-9627769-1-2

Designed by Lemieux Creative Enterprises

ATOMS, STARS AND MINDS:

SYNTHESIZING AN ELEMENTARY PARTICLE THAT COMPREHENDS ITSELF

Peter Bros

"...so as to enjoy in full the blissful shock, the enchantment and glee...on discovering... something in a scrambled picture--Find What the Sailor Has Hidden--that the finder cannot unsee once it has been seen."

SPEAK, MEMORY, Vladimir Nabokov

"You're right. I might be wrong."

Attributable as a frequent comment of H.L. Mencken.

THE COPERNICAN SERIES

AT THE GATES OF THE CITADEL
THE SUBJUGATION OF MODERN SCIENCE
How the scientific method generates facts, conceptually blinding us to reality

THE COOLING CONTINUUM
THE RISE AND FALL OF SPECIES ON EARTH
How life forms and evolves and why it becomes extinct

ATOMS, STARS AND MINDS
SYNTHESIZING AN ELEMENTARY PARTICLE THAT COMPREHENDS ITSELF
How matter forms and how it moves in space

THE MODEL MIND
HOW THE MIND MOVES MATTER
How the mind is constructed and how it forms, stores and recalls images

HOW THE WEATHER REALLY WORKS!
How air masses transport heat, and what transports the air masses

LIGHT
REPLACING THREE CENTURIES OF MISCONCEPTIONS
Unifying light, heat, electricity and magnetism with a single concept

INTRODUCTION

The job of pointing out that the Emperor had no clothes fell on an observer, who because of his youth, had not been trained to "see" the non-existent clothes.

The job becomes a little bit harder when the best tailors in the world have been admiring the Emperor's clothes for over three hundred years.

The Emperor's non-existent foundation garment is the belief that hypotheses can predict facts which, when found, will prove the hypotheses.

A hypothesis is a concept. Concepts are what the mind uses to organize facts. A hypothesis with a hole in it is just a more useful concept once the hole is filled with a fact.

The order in which facts are uncovered bears no relationship to the validity of a hypothesis.

Concepts are only useful in organizing facts. When they don't explain facts, they cease to be useful.

Thus, Galileo was wrong in asserting that hypotheses could be proven, and the scientific method, outside of its ability to prove the existence of facts by trial and error, is an erroneous procedure for purposes of analyzing reality to come up with a consistent picture of physical phenomena.

It deals with the existence or non-existence of facts.

It doesn't deal with concepts, which is what we have to use to organize the facts in a manner that provides us with comprehension.

Concepts can only be weighted by their ability to explain facts.

Newton put the non-existent golden outer garment over the non-existent foundation garment.

His mass/gravity concept rests on the assumption that objects move in a straight line unless a force acts upon those objects to make them move in other than a straight line.

This concept turns the universe on its ear.

Objects in reality do not move in a straight line until a force acts upon them to move them in other than a straight line.

Objects come to rest with respect to the existing forces acting upon them.

Newton's notion about clockmakers winding up the solar system at some time in the past is still with us in notions about the big bang that set everything in motion before our time.

Historic forces don't exist and aren't moving anything.

Assigning movement in the universe to historical forces is just a method of hiding a lack of comprehension about what current forces are operating to cause the movement.

Today, scientists spend much of their time analyzing black holes in space, conveniently marked with an "X" for easy notice, and the texture of the surface of the big bubble that resulted from some billions year old explosion.

Yet they can't provide an explanation why the Earth rotates on its axis.

Without such an explanation science literally doesn't have a clue why the sun comes up each morning.

Without comprehension of even the most obvious facts of life, why the sun rises, why objects move toward the surface of the Earth when they are released, even what light is, it is understandable that modern science would want to focus attention on areas that are totally outside the range of any observable physical phenomena.

While most practitioners of science will not even discuss the possibility the Emperor's old clothes are non-existent, the few that do feel a whole new set of clothes is a waste of time.

A new suit of clothes, the argument goes, will be just as non-existent as the old. Why discard what has served so well these hundreds of years? After all, over the years the non-existent golden threads have been encrusted with much jewelry that blinds the eye with its brilliance, and boggles the mind with its wealth.

These jewels were too hard won, and are too precious to replace with a new set of clothes.

It's too hard to start from scratch.

My answer is that it is better to have a pair of clothes that are cut to fit the emperor than a polyglot of dazzling jewels that serve only to hide the lack of foundation.

We need concepts that explain facts.

What we have are concepts that explain why other concepts don't fit the facts.

We are so far removed from the facts that we are dealing totally in the currency of hypothetical conclusions, hypotheses that have been accepted as fact, and which in turn have gone into the formation of other hypotheses which in turn are accepted as fact.

Because there is no predicted fact that human genius cannot eventually find, science is a hodgepodge of ad hoc hypotheses.

In pointing out that the world has fallen into a worse error than the error Copernicus confronted, I open myself to an endless ridicule of the type examined in detail in Chapter 17. Copernicus, in rearranging existing facts so that the sun, which obviously moved across the sky during the day and which therefore had to move around the Earth, was doing the opposite, decided to hold back on publication until he was on his death bed.

The mass/gravity mistake, which underpins virtually every reality we exist with, is more obvious than the sun going around the Earth.

I cannot count the number of times that I have been invited by the objective practitioners of science to jump out the window if I wanted proof of mass/gravity. It's slightly ahead of the number of times I have been asked to jump off a roof.

However, the existence of the attractive force provides a basis for its comprehension only by those that fail to comprehend the difference.

I have no intention of temporizing as a result of a general failure at comprehension.

And comprehension is the subject of this introduction.

How can just anybody without a sophisticated education in physics, astronomy, chemistry, biology, geology and philosophy understand a consistent picture of physical phenomena?

Aren't the multitude of facts too great? Don't the mathematical rigors involved eliminate all but certified geniuses?

The answer is a resounding no.

The only piece of equipment that we need to understand physical phenomena is the mind.

Physical phenomena are nothing but measurable facts.

The mind developed quite naturally to understand measurable facts.

Its tool to understand measurable facts is the concept.

The mind works on the basis of concepts.

Concepts are understandable.

Therefore, a consistent picture of physical phenomena

is understandable.

The fact that the picture of physical phenomena presented to us by modern science is not understandable should not undermine our confidence in our ability to understand a consistent picture.

The mass/gravity concept is an example of a concept that is simply not understandable. Watching an object fall to the surface of the Earth provides no understanding, has no meaning in and of itself.

It is a fact.

Saying that the object is falling to the Earth because it is a property of the object doing the falling is worse than providing no understanding because it provides an explanation with no understanding.

Saying that a rock falls to the surface of the Earth because it falls to the surface of the Earth is little better than saying it falls to the surface of the Earth because God makes it fall to the surface of the Earth.

While concepts are easy for the mind to understand, the process of conveying concepts is not so easy.

I developed the concepts related to the operation of physical phenomena to respond to the criticism that you can't criticize existing concepts unless you have better ones to replace them.

When it became apparent that objective science is probably more subjective than religion, many of its tenets being based on outright faith, I found myself exploring why reasonable people could profess objectivity while maintaining absolute subjectivity.

This led to the development of the concepts involved in how people think, how the brain itself operates.

Why does an entire planet of professedly modern, educated, and aware people slavishly believe something like the mass/gravity concept for which there is neither a theoretical nor experimental basis, a concept that is patently occult, the product of the mind of an alchemist derived three centuries ago before the development of physics, astronomy, chemistry, biology, geology and philosophy.

Having a consistent picture of physical phenomena and understanding how the mind comprehends the reality in which it exists still doesn't get me to first base without a method of communicating the underlying concepts.

If I have understandable concepts, and I have readers that are capable of understanding, then the only thing that stands between the reader and the concepts is understanding.

There are very few potential readers with an official knowledge of the facts involved.

Those that have knowledge of some of the facts, whose narrow interests intersect the fields of physics, astronomy, chemistry, biology, geology and philosophy that the concepts cover are committed to existing explanations for those facts and are therefore not among the group capable of understanding.

For those that can overcome their preconceived concepts, and for all others, obtaining the information about the facts is easy.

We are dealing with obvious facts, facts that most people take for granted: Light illuminates objects that fall to the surface of the Earth which rotates on its axis as it travels in an orbit around the source of the light.

There are currently no consistent concepts that provide a physical explanation for this simple statement.

The facts aren't a problem. Explaining them, specifically explaining them in a readable manner is the real problem.

One way to make something readable is to use short sentences and fill them with short words.

When sunlight is considered to be expanding, and the flows are diminishing as they expand, and their presence is proportional to how much they have diminished, and the effect that presence has is proportional to their presence, it is hard to put all this in a short sentence.

I have tried to use short sentences. The short sentences must be serial. I found out something testing short serial sentences. They can confuse comprehension. The confusion can be greater than a long series of clauses. Especially when the clauses are cemented into a single sentence.

I ended up alternating between the two approaches,

letting the meaning dictate the form.

Filling the sentences with short words is not as easy as it sounds, either.

For one thing, words have meaning, and long words usually convey the most meaning.

Second, we are dealing with an area that generates long words. Light is just part of the electromagnetic spectrum. Elementary and particle are both long words, and when used together, they are consecutive long words. When introduced in the fourth chapter, elementary is used eighty-three times and particle is used one-hundred times.

In like manner, chapters 15 and 16 introduce the concept of equilibrical matter to explain what the mind is and how it works. Equilibrical is used forty-seven times in chapter 15 and fifty-two times in chapter 16. However, elementary is still used sixty-five times in chapter 15 and fifty-four times in chapter 16.

My point is that, while there may be many long words in the text, they are the same long words.

Because there are so few long words, and the few that exist are used so often, I have found any dictionary of terms that I could construct to aid understanding to be unnecessary. The first few times you run into affinity propensity, at-rest motion, or the expanding field of electromagnetic emissions, it might give you pause.

However, these are not long words, but are the concepts that are used to build a consistent picture of physical phenomena.

The consistent picture that emerges owes its consistency to the fact that there are so few of the terms, and they are used so often.

I am not creating words to describe other words or effects.

I am using words to define concepts that can be used to describe the universe that we occupy.

I am not saying that it is going to be easy.

It's not.

But the learning curve is short, with the first hump being the mass/gravity concept.

Once you can accept the fact that mass/gravity is not a fact, but rather a concept, much of the explanation for physical phenomena that abounds today simply disappears because it is not designed to explain the physical phenomena we want to describe, the existence and movement of matter on Earth, in the solar system, and in the universe, and how it is that we can perceive both it and ourselves perceiving it.

That first hump is a brick wall for the sophisticated reader.

But for those of you that can make the short hop over the barrier of mass/gravity, I promise that the picture of physical reality that you find on the other side will amaze you. The vistas that open up will provide you with a new awareness of yourself and the universe we all occupy.

The universe is understandable, and it is understandable in physical terms.

And we are a part of that understanding!

1. PURPOSE

Not the purpose of life in the universe, nor the purpose of the universe. That will be discussed in the final chapters.

This purpose is the purpose of the book.

As can be seen, purpose can be understood, comprehended on any number of levels.

Just understanding, comprehension itself operates on more than one level.

We understand what we need to know to act in the physical reality of which our bodies are a part.

We want to go to the store. The trip requires the use of the car.

How does the car run?

We can take the key, put it in the ignition, and start the car.

This is the only thing we have to understand about the car to get it running.

What if you turn the key and nothing happens?

If you don't understand that there is an engine under the hood, then you are at a total loss. There is no understanding, and there is nothing to be done.

However, if you do know that there is an engine under the hood, and you know how to open the hood, then you have a different level of understanding.

You know that it is not the key that starts the car, but the engine.

You look under the hood and see that the wire that goes to the starter is disconnected.

You connect it, and you turn the key and start the motor.

Do you know why the motor started?

On one level, you have the technical know how that tells you that the key switch connects the starter to the battery, and therefore it will only work if it is connected.

But on another level, you have no idea why the starter responded to the key.

If you delved into it more deeply, you could understand that the battery produced electricity, and that electricity could be used to turn over the starter motor.

On this level, you could satisfy yourself about what was happening.

You could now say that you understood what made the key start the car. It connected the starter motor to the electrical source that gave it the energy to turn over.

But what is it about electricity that gives it the energy to start the motor.

You don't know, and science doesn't know.

With respect to electricity, current understanding is equivalent to an understanding about the car's operation when it's limited to turning the key.

Does this mean that there is an additional understanding of electricity, just as there was an additional understanding of the starting process of the car?

Certainly.

It's just that no one has bothered to obtain that understanding, being satisfied instead with measuring the effect of electricity rather than determining its nature.

SCIENCE IS KNOWLEDGE. A SCIENCE THAT CONTENTS ITSELF WITH A KNOWLEDGE OF HOW THINGS WORK, RATHER THAN AN UNDERSTANDING OF WHY THEY WORK IS BLIND SCIENCE.

What if we found that the car was wired, and everything was in mechanical order, but it still wouldn't start?

If we did not know that the car needs gas to operate, then we would have no understanding, and there would be nothing to be done.

However, if we did know that the car needed gas, then we would be operating on a different level of understanding.

If we put gas in the tank, we could start the car.

Do we know what it is about the gas that makes the car move?

We could take the engine apart, and use our technical abilities to see how the ignition of the gas in the cylinders was converted to the mechanical process that turned the wheels.

Would we know what is making the car move?

We might fool ourselves into thinking that we did. After all, knowing how the pistons rotate the crankshaft and then how that rotation is converted into the motion of the tires, this is pretty complex stuff.

But unfortunately for the advancement of knowledge, it isn't the piston, the drive shaft or the wheels that are moving the car.

It is the ignition and combustion of gasoline.

What is it about ignition and combustion that provides energy to move the car?

You don't know, and science doesn't know.

With respect to ignition and combustion, current understanding is equivalent to an understanding about the car's operation when it's limited to turning the key.

Does this mean that there is an additional understanding of ignition and combustion, just as there was an additional understanding of the starting process of the car?

Certainly.

It's just that no one has bothered to obtain that understanding, being satisfied instead with measuring the effect of ignition and combustion rather than determining its nature.

What if we don't have any gas?

If we know that gas is what we need, then we can find a way to get it. Because we didn't care too much about running out, we must have a way to get it. We know that there is a gas station down the hill from our house.

We can skip the business about knowing or not knowing about the car's braking system, and hop in, let the break off, and coast out the driveway and down the hill to get the gas.

MODERN SCIENCE DOES NOT ATTEMPT TO UNDERSTAND THE OBVIOUS. IT HAS CLUES ABOUT THE NATURE OF ELECTRICITY, COMBUSTION AND GRAVITY, BUT CONSIDERS UNDERSTANDING THEM BEYOND THE REALM OF SCIENTIFIC PURSUIT.

Do we know how the car is moving?

It's not the electricity in the starter motor, and it's not the gasoline combusting.

What is it that's making the car move from one point to the other?

We understand that the gas station is at a point lower than the point at which the car is and we know that objects fall.

We know that gravity makes objects fall.

So we know that gravity is making the car move.

But what is it about gravity that gives the car the energy to move?

You don't know, and science doesn't know.

With respect to gravity, current understanding is equivalent to an understanding about the car's operation when it's limited to turning the key.

Does this mean that there is an additional understanding of gravity, just as there was an additional understanding of the starting process of the car?

Certainly.

WHEN WE THINK THAT WE UNDERSTAND SOMETHING, WE DON'T QUESTION THAT UNDERSTANDING.

It's just that no one has bothered to obtain that understanding, being satisfied instead with measuring the effect of gravity rather than determining its nature.

If we wanted to delve into our understanding of the electric motor, we could substitute magnetism for electricity, ignition and combustion, and gravity and make the same set of statements.

We are in the same position with respect to these physical phenomena that we are when all we know about the car is how to turn the key in the ignition.

Because we fabricated the car, we can find somebody with the understanding that we don't have if the key doesn't work.

We can live with our lack of understanding.

With respect to the physical phenomena of electricity, ignition and combustion, and gravity, we have merely learned how to manipulate them.

In doing so, we have mistaken the ability to manipulate them with the ability to understand them.

When we deal with technology, we can always find someone that can understand what we don't because the technology came from somewhere.

When it comes to physical phenomena, we assume the same: When we don't understand something, then there must be someone somewhere that understands.

WE HAVE MISTAKEN OUR ABILITY TO MANIPULATE ELECTRICITY, COMBUSTION AND GRAVITY WITH AN ABILITY TO UNDERSTAND THEM.

When we think we understand something, or that if we don't, someone else understands it, we don't look for answers.

When it comes to understanding physical phenomena, we don't have the failure of a key to start the engine to jolt us into the realization that we don't understand something.

Without the feedback of failure, we have no way to mark our lack of understanding.

And if we don't have a way to mark our lack of understanding, then we won't make attempts to advance our understanding that would mark its failure.

Missed opportunity doesn't usually announce its presence.

If we think we know what gravity, magnetism, electricity, and ignition and combustion are, then we won't attempt to do things that would mark our failure to understand.

What we have accomplished with the knowledge of how to manipulate the physical phenomena in our environment would pale if we moved our understanding to the next level and achieved an understanding of the nature of electricity, ignition and combustion, and gravity. Even understanding what makes magnets behave the way they do would draw our accomplishments forward with breathtaking speed.

WHEN WE ARE NOT FORCED TO UNDERSTAND BY ADVERSE ELEMENTS IN THE ENVIRONMENT, ANY LEVEL OF UNDERSTANDING WILL SATISFY OUR CURIOSITY.

However, if we don't know we don't know because we think we know, then we will never try to know!

Galaxies are the primary structure of matter in the universe.

They are made up of stars such as our sun, and the stars that make them up inform us of their presence in the universe.

Matter defines the universe, it being actual existence

where there is no existence, non-existence.

Matter forms, ignites and combusts, emitting an electromagnetic spectrum. In its expanding electromagnetic spectrum, its combustion, we can see its location in nothing, non-existence.

It also moves in non-existence, and its movement tells us something about what is causing it to move.

We have precious few facts with respect to this matter, and we choose to ignore them all because we don't understand them.

What makes an object drop, what makes a match ignite and burn, what makes a light bulb burn, what makes an electric motor operate, all are such evident fixtures in our lives that the failure to mount a full scale effort to understand them would seem incredible.

IF WE DON'T KNOW WE DON'T KNOW, WE'LL NEVER TRY TO FIND OUT.

If it wasn't for the fact that we think we already understand them!

The sun comes up every morning?

Why?

Because the planet rotates on its axis.

Is this an answer?

No, it merely shows the physical relationship that makes it appear that the sun is rising.

What is the force that propels the planet to move around its axis?

Whatever it was, it happened five billion years ago.

Is this an answer?

Of course not.

The Earth speeds up and slows down.

In addition, things just don't spin forever, but rather come to rest with respect to the existing forces that are acting on them.

THE NOTION THAT PLANETS AND STARS MOVE UNLESS A FORCE ACTS ON THEM TO CHANGE THAT MOVEMENT IS ABSURD. IN REALITY, MATTER COMES TO REST WITH RESPECT TO THE EXISTING FORCES ACTING UPON IT.

We don't even know what makes the sun come up!

We also don't know why planets orbit, why they do so in the same direction, why they rotate on their axes in the same direction and why that direction is the same direction as the direction of rotation, or why they do so in the equatorial plane established by the sun's rotation.

How can the planets orbit the sun when the sun is moving through space at a speed estimated to be five to fifteen times the various speeds that the planets are moving, a hundred and forty miles per second as opposed to say twenty for the Earth?

And under these conditions, how can the moon still orbit the Earth at a speed independent of its motion around the sun, and through the galaxy, and perhaps with the galaxy's motion through the universe.

These are not dumb questions, but we have been satisfied by dumb answers, mumbo jumbo about historic forces setting everything in motion, and warps in space time continuums to keep them there, orderly chaos, and rolls down gravity wells.

If our purpose is to explain reality, then it is necessary to start with that reality, rather than making up a reality and then trying to explain it.

THE FAILURE OF MODERN SCIENCE TO UNDERSTAND FORCE EXTENDS TO ITS FAILURE TO UNDERSTAND HOW MATTER MOVES IN SPACE.

Granted it is safer to explain a reality that you have made up because then no one can question it.

But that doesn't get the reality explained, only ignored.

Reality presents itself in the form of physical matter in the universe, and the movement of that physical matter with respect to other physical matter.

If you don't understand it, and the nebulous conglomeration of fantasies that is perceived as modern science doesn't, then we are all really starting off on an equal basis.

From scratch!

We are all looking at the universe and asking the question, why is it the way we measure it to be?

Because matter is something that occupies nothing, it would seem that that something is made up of something.

In the absence of evidence to the contrary, we could use as a starting point the proposition that everything is made up of an elementary particle and that that elementary particle should be able to operate in such a manner that it will explain the universe as we measure it.

WHEN NO ONE UNDERSTANDS WHAT THE FORCE IS THAT MAKES THE SUN RISE EACH MORNING, WE ARE ALL ANALYZING PHYSICAL PHENOMENA FROM THE SAME STARTING POINT: IGNORANCE.

How do we know that such an imaginary elementary particle would really provide us with an understanding of physical phenomena?

What we do now is to take a phenomenon and make up an elementary particle that has the properties that would produce the phenomena we are using it to explain. Electricity is a moving charge, an electron if you will, light a photon, a unit of light, gravity a kind of unit of mass.

We wouldn't be doing anything different than we are already doing. We would just be using one to explain many rather than many to explain many.

If we make up a single elementary particle, what

would keep us from making it a chameleon particle, one that has such slippery properties that it could be used to explain anything?

If we look at reality to determine what the properties of the elementary particle are, and then make those properties fixed, unchangeable, require them to satisfy the physical phenomena that such a particle has to describe, then we could say that, while the elementary particle is imaginary, it allows us to understand how physical phenomena operate.

What this really raises, though, is the problem of what is understanding?

THE NATURE OF UNDERSTANDING IS ELUSIVE BECAUSE WE CANNOT KNOW WHAT WE DO NOT KNOW.

On the first day, God created the Earth provides a ready understanding of where the Earth came from, but it is a chameleon explanation, God being capable of all things.

God is a comprehensive explanation that explains nothing.

If there are different levels of understanding, then the question is, exactly what is understanding?

What is the physical process that is involved with our being able to form in our minds a representation of external physical reality, and then analyze that physical reality so that we can not only understand its operation, but understand that we are understanding its operation?

If we have synthesized an elementary particle from observation, and that elementary particle explains the physical phenomena in the universe, matter and the movement of matter, then it should also be able to explain how it is that we can perceive and comprehend that matter.

The same elementary particle that makes up matter should, with the same unchanging properties, result in our ability to form a representation of external physical reality in our minds, and allow us to know we are doing so.

WE CAN SAFEGUARD THE PROCESS OF UNDERSTANDING BY INDUCING FROM PHYSICAL PHENOMENA AN ELEMENTARY PARTICLE WITH SPECIFIC PROPERTIES AND THEN USING THAT PARTICLE TO EXPLAIN PHYSICAL PHENOMENA.

If we can accomplish this, we can be satisfied that we have moved our level of understanding to a level that would provide us with the ability to extend our range of survivability.

The concepts used to explain physical phenomena today are so weak that they have to be reinforced by the massive use of symbols to obscure understanding. Thus, incomprehensible mathematical systems and arguments are devised under the guise of safeguarding the process of understanding from the human frailty of misunderstanding.

Such a process allows the non-critical use of

hypotheses to generate fantasies unrestricted by reality.

One has only to look at the universe created by the unconditional belief in the mass/gravity hypothesis to see the result of unrestrained fantasy.

It is the purpose of this book to synthesize an elementary particle with two properties that can be used to describe the measurable universe in which we live, and the process that is involved in perceiving and understanding that universe.

I suspect that once minds realize that there is an understanding that lies beyond our current superficial understanding of physical phenomena, that the understanding will result in technological changes that might be thought capable under present concepts of science to prove the concepts involved in this book.

HYPOTHESES CANNOT BE PROVEN. THEY CAN ONLY BE EVALUATED.

Concepts cannot be proven.

Concepts are what the mind uses to understand measurable reality.

Measurable reality is what constitutes the facts.

Without concepts there is no way to comprehend, to understand those facts.

Thus, the only way we have of determining the validity of concepts is by their ability to explain the facts, and by their consistency with related concepts.

Therefore, it is not the purpose of this book to prove anything.

Its purpose is to provide the reader with a consistent picture of physical phenomena.

The word "science" has its root in knowledge rather than understanding.

Science today is knowledge, much of it manufactured by the inability to understand.

THE PURPOSE OF UNDERSTANDING IS TO CONSTRUCT A CONSISTENT PICTURE OF PHYSICAL PHENOMENA.

Obtaining a consistent picture of physical phenomena will lead to the understanding modern science has failed to provide.

2. METHODOLOGY

No one is going to get very far trying to change a tire with a screwdriver.

A pair of pliers won't do the job, either.

What you need is a lug wrench.

In the world of technology, not having the right tool is like having no tool at all.

So, too, in the attempt to construct a consistent picture of physical phenomena.

If you don't have the right tools, you aren't going to get very far.

If you are using the wrong tools, and you think that you are using the right tools, the result could be catastrophic.

Remember, we aren't dealing here with a key that normally starts the car. We don't have a signal that we are failing in our effort to come up with a consistent picture of physical reality.

Unless, of course, the picture we come up with is so inconsistent as to be incomprehensible.

Which is what we have today.

Almost five hundred years after Copernicus, about the only thing that we really know about the universe is that the Earth travels around the sun. Because we didn't just sail off the surface of the Earth, no one could believe this.

It took Galileo's demonstration of the situation using the moons of Jupiter as a representation of the solar system to drive the fact home.

Galileo did this some hundred years after Copernicus first rearranged the concept that put the Earth somewhere other than at the center of the universe.

However, this still didn't explain why we weren't flying off the surface of the Earth if the Earth was actually moving someplace.

Newton solved this conceptually after another hundred years by hypothesizing that we didn't fall off the Earth because there was a force holding us to the surface of the Earth, and the force was a property of the matter that made up the Earth. We didn't fall off because the force of Mother Earth held us to her bosom.

THERE IS NO CONCEPTUAL OR EXPERIMENTAL EVIDENCE TO SUPPORT THE NOTION OF MASS/GRAVITY.

It was such a relief to know that we weren't all going to be hurtled off into the deep dark coldness of space, that we bought this little piece of fancy lock, stock and barrel, and have since used it to create a universe in which the things that are real, the movement of matter in space, are ignored, and the things that have no existence, space and time, are quantified and become reality.

In the three hundred years since, Newton's unfortunate assumptions led to a belief stronger than the belief in any deity, becoming itself a deity, whose presence is so all pervasive that to think in its absence today is an impossibility.

The source of the confusion, the assumption that allows us to believe that our concepts are in reality facts, can be laid at Galileo's doorstep.

Galileo created a tool to be used in the effort to construct a consistent picture of physical reality that was, and is not appropriate to the task.

This is the scientific method, and it is probably the key that has kept the level of understanding of physical reality in the dark ages for almost half of a millennium.

> **THE NOTION THAT PREDICTIVE FACTS CAN PROVE HYPOTHESES CREATES SCIENCE BY CONSENSUS IN WHICH CONCEPTS NEED ONLY BE CONSISTENT WITH CONSENSUS RATHER THAN WITH EACH OTHER.**

Unlike a key in a car, it is a key that always works, and therefore leads us to believe that we are comprehending the universe, riding breezily down the road of knowledge, when in fact we are still in the driveway, making up magic pictures in our minds that have no basis in reality.

The scientific method is simple in its conception because what it can do is deceptively simple.

If somebody makes a statement, say that two objects of different weight fall at different rates, then the scientific method says you can find out the truth of the statement, where this is in fact a situation that exists in physical reality, simply by taking two objects of different weights, dropping them, and measuring the interval between the time they are dropped and the time that they hit the ground. If the distance (space) is identical for both objects, and the interval (time) is the same, then the statement, that two objects of different weights fall at different rates is not a statement that reflects physical reality.

What the scientific method does is allows us to measure conditions in physical reality so that when we reconstruct that physical reality in our minds for analysis, we know that our mental constructs are at least limited by what we can measure in physical reality.

We may not know what causes the objects to have different weights, another physical reality altogether, but we know that whatever it is about the force that is making them fall, and whatever it is about the basic structures that make up the matter, it has different effects when the matter is moving freely within the force than when it is moving against the force, the latter situation being what gives the measure of weight.

The scientific method, as a tool in the effort to construct a consistent picture of physical reality, gives us the dimensions of that reality.

It does nothing to provide us with an understanding of how that reality might work.

We can measure the rate at which an object falls, but taking the measure of that rate, or knowing that the sun or the moon or the planet Jupiter might also make objects fall toward its surface tells us nothing about why the object falls.

This information is not directly accessible to our ability to measure. We can't measure how the force is generated, how it moves through space, or how it acts on matter to make it move in space.

If we can't get the information by measuring it, we have to find another way to obtain it.

We can't use the method we designed to measure reality where that reality is not measurable.

> **THE CONCEPT OF MASS AS A PROPERTY OF MATTER IS OCCULT, A MODERN DAY EXPRESSION OF FAITH IN WOOD NYMPHS.**

But that's exactly what Galileo said we could do, that's what Newton did, and this misuse of a tool has driven science into a failure not only to understand physical reality, but a refusal to confront it.

Galileo made a hypothesis about measurable reality that could be proven in physical reality simply by measurement. He hypothesized the optimum angle a projectile could gain distance by rolling steel balls down an inclined plane in the laboratory. He later showed the laboratory hypothesis to be a fact in physical reality.

Galileo then made the deadly mistake of jumping to the conclusion that any hypothesis about physical reality could likewise be proven!

His thought was that making a hypothesis was to make a conclusion about the operation of facts. Given certain facts about measurable reality, the hypothesis should be able to predict how other facts operated in that physical reality.

If the predicted facts were found, then the hypothesis must be correct.

This is erroneous.

HYPOTHESES CAN BE EVALUATED BY THE FACTS THAT THEY EXPLAIN, NOT BY THE FACTS THAT THEY PREDICT. FINDING A PREDICTED FACT MERELY INCREASES THE USEFULNESS OF A HYPOTHESIS.

Facts exist independently of hypotheses, which are merely concepts created to explain facts. If a concept explains a non-measured fact, subsequently measuring the fact doesn't turn the hypothesis into a fact. It just makes a concept that explains one more fact!

The only procedure that can prove a hypothesis is one that shows the hypothesis to be a measurable fact.

The hypothesis that gravity is a property of and proportional to something called mass is a non-provable hypothesis.

In fact, Newton's units of mass, which he was using in the sense of size with respect to the size of the moon as it related to the size of the Earth were quickly abandoned when the inverse square law didn't work for anything else in the universe, and eventually not even the moon, when it was found that it wasn't made of green cheese or had a sunken backside (to account for its lack of mass).

By shifting the focus from size, which didn't work out, to density, which was non-measurable, a quantity was substituted which could always be determined from the inverse square formula.

The result was the ultimate expression of the scientific method, a hypothesis that could never be disproven!

NEWTON USED A CONCEPT OF MASS TO PROVE THE INVERSE SQUARE LAW. THE INVERSE SQUARE LAW WAS THEN USED TO DETERMINE MASS.

The inverse square law has nothing to do with the fact that the positions of the planets are generally predictable. They have always been predictable because the solar system operates on a repetitive basis.

Taking sight measurements, saying that they are derived from a universal law of gravitation where gravity is a property of matter, and then covering the failure of the law by factoring in a totally non-measurable assumption (the make up of the cores of the planets) is not science, but chicanery, pure and simple, sleight of hand that has no place in the attempt to come up with a consistent explanation for physical phenomena.

We start out with measurable conditions, derive a

measurable law, and when the measurable law doesn't work for the measurable conditions, we create a non-measurable term in the equation to render the measurable law disprovable.

Newton started out to prove the inverse square law, and when it didn't work, we said, well, it's true, we can't measure the insides of planets, so the law proves the make up of the insides of the planets.

What makes that whole Galileo to Newton process even more discouraging is the fact that objects do fall according to an inverse square law. That was known even before Galileo. All that is necessary to find out about this physical relationship is to drop an object and measure its rate of fall. That rate increases in a manner that shows that the force acting on the object is increasing at a rate inversely proportional to the distance over which it falls.

IF THE SCIENTIFIC METHOD CAN ONLY DISPROVE FACTS, THEN HYPOTHESES THAT CAN'T BE DISPROVEN BECOME FACTS.

This was known well before Galileo convinced everyone you could turn speculation into fact by finding a predicted fact (something that is always achievable) and Newton clothed his alchemical notions of force and matter into a bunch of Latin mumbo jumbo that has led attempts to understand physical reality into an abyss of unreason.

If the scientific methodology is not a valid procedure to

prove hypotheses, then what is?

Hypotheses are concepts unless they have the possibility of becoming facts. Most facts are settled today. We know at what rate objects fall. We have a pretty good idea of the dimensions of the solar system. We're not too good with other dimensions, space, or the interval of events, time, but we can examine the electromagnetic emissions of stars and galaxies and conclude a little bit about the universe that we find ourselves in.

But the only things we have to use to understand that universe are concepts.

If concepts are not provable, then what guarantee do we have that the consistent picture we create of external reality has anything at all to do with that external reality?

How do we avoid placing the Earth at the center of the universe, or enshrining gravity in mass?

How do we know that we are going to be understanding how the universe operates so that we can move forward in that reality with comprehension?

We can know by finding a way to physically describe how gravity might work, physically describe what electricity is, physically describe what is happening when something is undergoing combustion. We should be able to physically describe the universe like the way we can physically describe the way pistons convert force to move the wheels of a car.

But how do we know what we are doing is going to

move our comprehension forward?

When we try to use concepts to understand the universe, how do we know those concepts reflect reality?

The answer lies in the attempt.

The attempt is to find a consistent picture of physical reality.

If we end up traveling down a whole series of pathways, none of which are consistent with any of the others, then we are not establishing a consistent picture of physical reality.

A SCIENCE THAT MAKES NO ATTEMPT TO UNDERSTAND THE MOVEMENT OF MATTER AND THE EXISTING FORCES IN WHICH MATTER MOVES IS A CHARADE RATHER THAN A SCIENCE.

It is not consistent to know that objects come to rest with respect to the forces that act on them, a fact that can be demonstrated time and again by simply falling off a roof, and then saying that the solar system, nay the entire universe, was set in motion at some distant time in the past and is moving as a result of those historical forces.

Things don't move as a consequence of historical forces.

They move as a consequence of current forces.

The forces that cause the planets to rotate, that cause them to revolve around the sun, that causes the sun to hurtle through space at seven times the rate of the Earth (about one hundred and forty miles per second), that causes the Earth to move around the sun are not historical forces.

They are current forces, and a scientific establishment that does not make its primary activity the attempt to determine exactly what those forces are is leaching off society, taking its resources with the promise of returning understanding, and then obfuscating that understanding to hide from society its own failure to comprehend.

We will see in later chapters that the mind operates on the basis of comparison.

Combustion creates flows of the electromagnetic spectrum (light). These flows expand and bounce off objects in reality. When they do so, their rate of expansion is altered by the angles the objects in reality make with the source of the light and the eye.

Each of these angles is unique which makes each of the flows unique. The angle the flow makes with the object in reality determines the presence of the flow, and it is this information that gives us our knowledge of external reality.

The eye determines the presence of each of these flows by the information obtained from the angles and in doing so constructs a picture of that reality.

We will see in later chapters how the microstructure that represents this information, made out of the same elementary particle that will be used to create atoms, and the fields of force in which they move, will move into the structure of the mind, be reconverted back into the streams it represents, and in turn be reconstructed into a microstructure for storage and use in comparison with external reality.

For purposes of determining some sort of methodology to deal with concepts, it is necessary to understand that the mind operates on the function of reconstructing external reality and comparing that external reality with internal reality.

PHYSICAL PHENOMENA HAVE MORE THAN ONE CHARACTERISTIC. DISPARATE PHYSICAL PHENOMENA HAVE IDENTICAL MEASURABLE CHARACTERISTICS. THE VALIDITY OF THE CONCEPTS WE DERIVE TO ADVANCE OUR UNDERSTANDING CAN BE CHECKED BY THESE IDENTICAL MEASUREMENTS.

If facts are linked in external reality, then that linkage will control the relationship of the facts in internal reality.

We first have to limit our internal world by measurable reality.

We then have to find facts in external reality that are linked together, and take these links, and in turn link them together by their identical facts.

Facts that always appear with other facts are factual congruities.

Light expands and diminishes!

Objects fall and accelerate!

The process of recall and comparison which produces understanding can be structured to external reality by the process of linking these factual congruities in physical reality.

It is this process that allows us to develop concepts for understanding physical reality with a high degree of certainty that they are based in the real world.

3. FACTUAL CONGRUITIES

Two plus two equals four.

What is this?

Mathematical equations are useful in telling us something that we need to know about physical reality.

They do not provide us with comprehension about that reality itself.

Unless we have a purpose for counting something, equations derived from the need for counting serve no purpose in nature.

Most of the necessity for counting came from commerce. When the economy passed bartering, where the determination of value was made by the simple desire or need to possess, the amount of possessions turned over to others for indirect barter through the use of a medium of exchange gave rise to the necessity of counting.

If I have two bushels of corn, and you have two bushels of corn, and we put them together in the hold of a ship, we have to have some sort of system to know how much each of us has.

From myriad transactions such as this, a system of mathematics builds up which allows us to engage in commerce.

But the counting systems, outside of our needs for transactions, are quite useless. It is nice for poets who have learned the counting system to lie on the beach and wonder how many grains of sand there are on the beach, or how many points of light there are in the sky, but such efforts serve no purpose other than the aesthetic desire to use the mind for purposes other than those designed to reflect reality.

Other mathematical systems grew up out of necessity. If we want to know how tall a building is, and we can't measure it directly, we can devise methods, formulae, that will give us the answer if we have other measurements.

MATHEMATICAL SYSTEMS ALLOW US TO PREDICT THE AFFECT OF FORCE ON MATTER BECAUSE OF THE FEEDBACK INVOLVED IN ACTUALLY TESTING THE EFFECT OF THE FORCE ON MATTER IN REALITY.

As we get more sophisticated with our technology, our mathematical systems become more sophisticated, allowing us to determine the stresses manufactured substances will endure, the amount of lift certain flows will create over sculptured surfaces, the amount of radiation necessary to kill a cancer without killing the patient.

Mathematical systems allow us to manipulate physical reality.

They do not comprehend physical reality, nor do they provide comprehension of physical reality.

Measuring the dimensions of a ball does not provide information about the structure of the ball.

Today, when mathematical systems are constructed without reference to physical reality, and then physical reality is conceptually structured to conform to them, there is much marvel at how the human genius can create the solutions to problems before the problems are even known.

This devotion to mathematics at the expense of reality is similar to the devotion of Ptolemaic preservationists in the face of the Copernican conclusion.

Unlike the idea that the Earth was flat, an idea that was never believed by people who were spending time trying to order the facts of the universe around central concepts, the idea that the Earth was the center of the universe around which the sun traveled was as fixed in its day as the mass/gravity concept is today.

THE SCIENTIFIC METHOD WORKS IN THE CONTEXT OF TECHNOLOGY BECAUSE ITS ABILITY TO DISPROVE IS INHERENT IN THE NATURE OF THE TRAIL AND ERROR INVOLVED IN TECHNOLOGICAL ADVANCEMENT.

It was quite obvious that the sun traveled around the Earth. One only had to walk out the door every morning and watch it do so. You could even use mathematics to determine the rate it traveled.

When the orbits of the other objects in the heavens came to be analyzed, however, their motion conflicted with the idea that the sun traveled around the Earth.

The starting point was the Earth at the center of a number of concentric circles each of which represented the orbit of a known planet, the moon or the sun, much like the picture of the solar system as we see it today, only with the Earth at the center.

As the ability to measure improved, this picture was proven to be wrong.

However, instead of changing the basic concept that held the Earth at the center of the universe, when the motion disagreed with the concept, the centers of the various circles were altered. Just as we can visualize our moon moving around the Earth, or Jupiter's moons moving around it in a number of concentric circles, which in turn travel in concentric circles around the sun, planets were made to move in concentric circles around hypothetical planets that were moving in concentric circles around the sun that was moving in a concentric circle around the Earth.

Every time that a new measurement didn't agree with the existing concentric circles, the existing concentric circles were altered and reformed to bring them into agreement with the observation.

However, the observations could not be brought into agreement with reality because no matter what the observation was, it couldn't disagree with the concept that the Earth was at the center of the primary set of concentric circles.

That fact, a concept, could not be altered regardless of the observation.

As the centuries passed, concentric circles grew, shifted and changed, hovering around seventy, with terms created to describe what was manufactured in the mind. Ptolemey's epicycles, deferents, and equants foreshadowed our mass.

Copernicus, using the principle that the simplest explanation was the best explanation, reduced the number of circles by positioning the sun rather than the Earth at the center of the solar system.

WHEN A CONCEPT IS ACCEPTED AS FACT, FACTS THAT DISAGREE WITH THE CONCEPT ARE IGNORED OR EXPLAINED WITH THE USE OF AD HOC THEORIES THAT DON'T DISAGREE WITH THE ACCEPTED CONCEPT.

The mathematics that were used to obtain the measurements, and then the mathematical systems that were created to explain the concentric circles within concentric circles within concentric circles did nothing to increase the understanding of physical reality.

As long as the concepts dealing with that physical reality were not examined, questioned and altered, all the mathematics in the world did nothing more than confuse the understanding of physical reality.

Mathematics can be used to test physical reality, give physical reality its dimensions, but they can't give physical reality meaning.

It is the human mind that gives meaning.

To believe that mathematical systems provide a higher meaning than the mind, as science does today, is to consign the mind to the dark days before Copernicus, spiraling deeper and deeper into incomprehensible inconsistency.

Copernicus used Ockham's razor to justify the conceptual shift. Ockham, who preceded Copernicus by several hundred years, realized that the mind is unbridled in its ability to explain physical reality, and that therefore, that explanation which explains the most in the simplest manner must be the explanation that prevails.

CONCEPTS, RATHER THAN MATHEMATICS ARE THE TOOLS OF COMPREHENSION.

Ockham's principle foundered from the beginning, and is still not a viable methodology today because a deity can be used to explain the most in the simplest fashion.

The real problem is the definition of the word "explain".

We have seen that there are several levels of understanding and that the level is limited by the explanation we are satisfied in accepting.

Pre-Copernicans were content with the level of explanation that put the Earth at the center of the universe. They occupied themselves with drawing their endless maze of intricate concentric circles.

We are content with the level of understanding that imprisons gravity out of reach in matter and occupy ourselves with drawing endless concentric circles with the invisible ink of time embedded in the touchless fabric of space.

The pre-Copernicans were not about to send a man to the moon, so their concentric circles really were abstract. They did not understand the relationship between knowledge and survival, their survival being dependent on salvation rather than the alteration of physical reality.

Our concentric circles matter!

WHEN AN ERRONEOUS PICTURE OF REALITY DOES NOT AFFECT SURVIVAL, ANY PICTURE IS ACCEPTABLE.

It's hard to see how the pre-Copernicans could have tested the concept involved in the change of concept to a heliocentric solar system.

Mathematics was just as useless to them in their attempt to understand the universe as it is to us.

If "two plus two equals four" won't do it for us, what will?

We have to know how the mind works to protect the mind from itself.

The mind is constructed in such a manner that it can only construct a single picture of reality at a time. Some people may be adept at changing the picture in the mind so fast that they think they have two pictures in the mind simultaneously.

But we shall see that in constructing the mind out of the elementary particle that will be used to explain physical reality and the mind, the mind can only reconstruct a single picture at a single time.

However, because it operates on the basis of comparison, it can alter the picture it holds by recall.

And because it re-encodes the pictures it forms, it can re-encode the altered picture.

FIVE HUNDRED YEARS AGO, OUR PICTURE OF THE SOLAR SYSTEM WAS IRRELEVANT TO SURVIVAL. TODAY, OUR SURVIVAL DEPENDS ON HAVING AN ACCURATE PICTURE OF REALITY.

This process is what allows us to adapt to changing conditions, and after we have adapted to look back and have no recollection how we existed in the conditions before the change.

We constantly live in the present because we are constantly reconstructing the past by re-encoding it with the present.

The process allows us to extend our comprehension of physical reality by taking one thing we know about that physical reality, combining it with something else in that physical reality, and re-encoding it so that when we recall it we have a picture of physical reality that didn't exist before we combined the two into a new picture of physical reality.

This is how we comprehend physical reality.

Because we can combine concepts, and mix them with facts, our process of comprehending can lead us away from comprehending physical reality.

However, if we base our comprehension on facts that exist together, then we anchor ourselves in physical reality for these factual congruities.

There was no understanding of factual congruities in Copernicus's time, a growing awareness of them in Galileo's time, a disavowal of them by Newtonian mechanics, and no understanding of them at all in our time, our comprehension having moved so far from physical reality as to make physical reality unimportant.

WHEN TWO FACTS IN REALITY ALWAYS EXIST TOGETHER, WE COMPREHEND SOMETHING ABOUT PHYSICAL REALITY THAT IS NOT ERRONEOUS.

There are only a few factual congruities, and they are both the key to understanding physical reality and constructing a consistent picture of physical phenomena, and the way to check the reality of that picture and the underlying understanding it conveys.

It is important that we be clear in our concepts by being clear in how we describe the physical reality our concepts are to encompass.

What does an object do when we release it from our grasp?

We are used to thinking of it as falling toward the Earth.

It is this way of describing what is happening that gives the mass/gravity concept its factual nature, its appearance of reality.

What is happening when we release an object from our grasp is that it begins to move.

By holding it in our hands, we are restraining it from moving.

WE DO NOT DROP AN OBJECT AND LET IT FALL. WE RELEASE AN OBJECT SO THAT IT CAN MOVE IN RESPONSE TO EXISTING FORCES.

Today's concepts lead us to believe that somehow when the object is moved against a gravitational field, it takes on energy, stores it up in some mysterious way so that it is just waiting there to expend it. This is the type of concept that put the energy in planets to move them a putative five billion years ago and allows them to continue to expend it.

Today's concepts lead nowhere.

If we are holding an object, and someone removed the Earth, the object would no longer have any energy to move, so that it is not storing anything.

If the object is not storing anything, then we are restraining its movement, and when we let go, it begins to move.

The factual congruity here is how it moves.

We can use mathematics to measure the dimensions of external reality. One of those dimensions is how matter moves.

We have no idea at this point what is making it move. We just know that we can use our muscle power to restrain its movement, and we know that once we release our restraint it will begin to move.

By measuring that movement, we know that the movement is an accelerating movement. This is to say that while it is moving, concept number one, it is accelerating, concept number two.

Because this situation always exists it is a factual congruity, two facts that always exist together, that are congruous.

There must be the force we term gravity to make the object move, but the factual congruity is how matter moves with respect to that force.

Because the mind can only understand one concept at a time, can only create one picture of physical reality at any particular moment, and the factual congruity presents us with two concepts, a moving object and an object moving at an increasing rate, we have to combine the two concepts into a single concept to deal with it.

This is why factual congruities are so elusive.

We must first visualize a moving object, and then we must have a way to recall an object moving at an increasing rate, say an accelerating car, and we can then re-encode the new concept as it relates to the movement of an object in a field of gravity.

The measurement that is derived to apply to the concept of an object released in a gravitational field is that the object accelerates in that field as if the field were increasing in strength inversely with the square of the distance over which the object is moving.

Which immediately raises the second factual congruity.

WHEN WE AGREE ON SOMETHING BECAUSE WE HAVE NO EXPLANATION FOR IT, WE CALL IT A LAW.

If something is accelerating, it is doing so at a rate.

In the first factual congruity, the two facts were movement and acceleration.

These two facts always appear together with respect to an object released in a field, so we don't need a measuring tape to validate their existence in external reality.

They are locked together.

Now, we take one of the facts in that factual congruity and connect it to another fact to find a second factual congruity.

The second fact in this factual congruity is what is known as the inverse square law. However, laws are devices used to substitute conclusions for understanding. Therefore, it is really just a measurement.

Objects released into a gravitational field accelerate at a rate inversely with the square of the distance over which the movement occurs.

To arrive at this factual congruity, we have to recall the re-encoded concept that provides us with a picture of a moving object accelerating and give it a fixed rate of acceleration determined by the distance traveled.

We can see that by the use of factual congruities and the most rudimentary of measuring ability, we have kept ourselves rooted firmly in physical reality.

THE UNRESTRAINED MOVEMENT OF MATTER IS ONE OF THE MOST OBVIOUS FACTUAL CONGRUITIES IN PHYSICAL REALITY.

We don't know what it is that is making the objects move, but we now know a lot about it: It's something, it makes objects move, it makes them accelerate as they do so, and it makes them accelerate at a rate that is inverse with the square of the distance over which the movement occurs.

We can encode this into a single factual congruity, and we end up with one of the factual congruities in external reality, one that we know reflects reality even though it is derived from the use of concepts that are not subject to being proven.

Unrestrained objects move in physical reality in

accordance with a force that increases inversely with the square of the distance over which the objects move.

What other factual congruities are there? The factual congruity dealing with an object released into a field was measurable by the simple use of the "one plus one equals two" process developed to make commercial transactions possible. Is there a factual congruity inherent in the measurements derived from a physical process?

When we strike a match, we think in terms of energy stored in the sulfur head that is released upon ignition and combustion of the match. We think of the matter as one thing, and the light as another.

If we concentrate on the light, we can see a second factual congruity whose measurement is inherent in the physical process.

The first fact that goes into the factual congruity is that the light expands. It moves away from the match in all directions. The second fact is that something that moves away from a point, the match, in all directions forms a sphere.

If we think first of the light moving away from the center of its source, the match, and then recall a balloon blowing up, we can re-encode the factual congruity, light moving away from the ignited match like a luminous balloon blowing up.

The new concept allows us to think of light moving out from the sun in the form of an expanding sphere, the two facts being the expansion and the sphere. Because light is moving away from the sun at each moment, at each moment a new expanding sphere of light is being created and these expanding spheres extend from the surface of the sun to as far as the light from the sun can be seen.

Once we re-encode this concept, which again we know is based in physical reality regardless of the fact that it is a concept, then we can use the fact of an expanding sphere to discover another factual congruity.

Shapes have inherent relationships in physical reality, and the relationship of the areas of the surfaces of spheres is determined by their radii, the distance from the surface to the center.

> **LIGHT, HEAT, THE ELECTROMAGNETIC SPECTRUM IS ANOTHER OF THE MOST OBVIOUS FACTUAL CONGRUITIES IN PHYSICAL REALITY.**

While we can't actually measure what happens to light when it travels from one place to another, we know that it diminishes.

Thus, if we hold in our mind the re-encoded concept of an expanding sphere of light, and then recall how light can become dim over distance, we can re-encode the new factual congruity, that light diminishes as it expands over the surface of an expanding sphere into a single concept.

But as soon as we know it diminishes, we are faced with the question of rate again so that we can take the new fact, light diminishing over the surface of an expanding sphere, and combine it with another fact, the rate at which it diminishes, to create still another factual congruity.

Because we know that the surfaces of the spheres over which light is expanding are related by the square of their radii, the resulting factual congruity that can then be re-encoded into a single concept is that light expands over the surface of an expanding sphere, and diminishes as it does so inversely with the square of its distance from the source of the light, the center of the sphere.

> **WHEN WE HAVE TWO FACTUAL CONGRUITIES WITH A COMMON MEASUREMENT, WE CAN CONNECT THE TWO WITH A REASONABLE CERTAINTY THAT THE CONNECTION EXISTS IN PHYSICAL REALITY.**

The present concept is that it does so uniformly, and it is this concept that has hidden the mechanical cause of an object moving from one place to the next when it is released into a field, but we will get to that later.

For now, we can create still another factual congruity using the fact that an object accelerates in a field at a rate inversely with its distance of travel and the fact that the expanding sphere of light decreases at a rate inversely with its distance from the source to create the concept that an object moves toward the source of an expanding field of light at the same rate that the light diminishes as it expands.

In other words, if we can find a mechanism in the process of the expansion of light, it would work in precisely the same manner as the force accelerating an object moving freely within it.

We can locate gravity in the expanding electromagnetic spectrum rather than in the matter that produces the expanding electromagnetic spectrum.

There is one additional factual congruity, simple in concept but extensive in application. Electricity is considered to be a moving charge. As it moves along a path it creates an inductance. These inductances are considered to be a force with direction. The direction of the force is determined by the ubiquitous right hand rule, in that if the thumb of the right hand is pointed in the direction of the moving charge, then the fingers curl in the direction of the force of inductance.

The factual congruity is that inductances are always proportional to flow. If the flows increase, the inductances increase. If the flows decrease, the inductances decrease. If there are flows, there are inductances. If there aren't flows, there aren't inductances.

As we shall see, this factual congruity is the most prevalent factual congruity in the universe.

And finally, before we establish our hypothetical congruity, a word of caution on something extremely important that has been invisibly accepted as a factual congruity, which is in fact two separate but far reaching facts.

Objects have differing weights. Weights are thought to

be the result of particles in the matter that have weight, with matter with more weight having more particles with weight.

However, weight is nothing more than the relative effort it takes to move atoms against a field of force. If atoms are released into the field, they move in that field at a uniform rate and the bindings that hold the atoms together exert no effect on the movement.

However, if atoms are moved against the field, then the difference in the effort it takes to move different atoms depends on how many atoms, and units of atoms are held together by their bindings.

THERE ARE IN PHYSICAL REALITY TWO OBVIOUS FACTUAL CONGRUITIES, AND ONE PREVALENT ONE. MASS, MOMENTUM AND INERTIA, CONCEPTS OF MATTER MOVING WITHIN AN EXISTING FIELD OF FORCE ARE NOT AMONG THEM.

Momentum is considered to be a factual congruity composed of weight and movement, with mass increasing with momentum.

In reality, momentum and weight are the product of the way in which atoms, and the units that make them up are bound together.

The choice of whether to have a hundred pounds of iron or a hundred pounds of feathers dropped on you would depend to a large extent on whether the iron was in one once floating flakes and the feathers were bound together in a single bale.

4. THE HYPOTHETICAL CONGRUITY

There has to be an elementary particle that is basic to everything.

While this concept stretches back through the millennia, it was during the scientific diversification that followed Newtonianism that particles were assigned to the properties they were designed to explain and then the properties to the particles.

This proliferation of electrons, neutrons, protons, gravitons, magnetrons, photons, and thermions was so successful that when the twentieth century rolled around the number was multiplied to the point where incomprehensibility made the tillers of the learned fields impervious to criticism, even from their peers.

IF A SINGLE PARTICLE CAN BE USED TO PROVIDE A CONSISTENT PICTURE OF PHYSICAL PHENOMENA, THEN IT IS UNNECESSARY TO CREATE A PARTICLE TO EXPLAIN EACH EFFECT.

We are setting out to explain with a single elementary particle the galaxies, and the atoms that make them up, together with the mind that comprehends them. The single elementary particle will end up doing the same thing that all the other particles do, at least those that explain actual rather than imagined physical phenomena, so we will start by defining that elementary particle.

The particle we are going to synthesize is close to the

concept of an electron combined with a photon, with the effect of a thermion. With the thermion being the least defined and used of the three, I have tended to use that designation. However, for purposes of this synthesis, I will refer to the elementary particle as the elementary particle.

Many of us have the problem of thinking of the universe in terms of our man made measurements. If we measure to the end of the universe, what is outside of the end.

WE CREATE MEASUREMENTS TO ALLOW US TO DESCRIBE MATTER IN REALITY. THE MEASUREMENTS THEMSELVES, SPACE AND TIME ARE NOT THEMSELVES MEASURABLE QUANTITIES.

We created measurements to measure distances, spatial and time, between points and events, and the resulting quantification of space and time are useful only when there are points and events.

Without matter, there are no points and no events, so that there is nothing to measure.

If there is nothing to measure, then there is simply that, nothing.

We only have trouble conceiving of nothing when we try to do so with respect to something.

Nothing is non-existence, the absence of matter.

It simply doesn't exist.

There is not even an "it".

We don't have to deal with non-existence.

It is existence we have to deal with.

Existence is matter in non-existence.

Matter is made up of elementary particles.

What, then is the basic requirement of the elementary particle?

How big is it?

If there is nothing, non-existence, and an elementary particle for whatever reason, comes into existence, it would only have to be large enough to define the non-existence in which it found itself.

These are not abstract questions. While today's concepts about the universe would prefer to have matter in existence in the dualities of before and after, occupying space time that is both here and there and before and after, and which bends around like some sort of universal uroboros to bite its own beginning, they don't get to the crux of the situation.

Matter exists, so it had to come into existence somehow. We can speculate about that "somehow" later, but because matter exists, we have to accept it as coming into existence, and because elementary particles make up matter, we have to accept a single elementary particle coming into existence in nothingness, non-existence, and it is only then that we have to determine just how big that elementary particle is.

And because there was nothing before it existed, it has to be no bigger than is necessary to define that nothingness.

Today's concept of the elementary particle is derived from the most unfortunate analogy ever made, the assumption that light behaves in the same manner that water waves behave as a result of gravitational interaction. This analogy ranks with mass/gravity as the sower of the confusion that reigns in the attempt to comprehend the universe and results in an elementary particle, pick an electron or a photon, that is a wave particle, a fluctuating point of energy, an approximation of the averages that give it existence, a mathematical equation, or the end of a one dimensional string.

All of this is quite incomprehensible!

We want an elementary particle that we can use to describe physical phenomena so that we can come up with a consistent picture of reality.

Our particle will actually be an existing particle, actual matter, and the matter will be just enough matter to exist, and by its existence to define non-existence.

If we put ourselves as non-material observers in non-existence, what would we see when that particle came into existence?

Probably just a distortion of non-existence.

We wouldn't have anything to compare the particle

with, so we would have no knowledge of its size or its motion. We have to assume that if it is only big enough to define non-existence, it must be spherical in shape, a tiny dot that makes us aware of the non-existence that stretches out from it in every direction.

But we couldn't derive a measurement because there is only the single point. Before there can be a measurement, there must be a second point.

So let's create a second elementary particle, and place it a distance from the first elementary particle in such a way that the two maintain that distance exactly.

We have not only created a second elementary particle, we have created a measurement in physical reality, what we can now define as the existence of matter in non-existence.

IF WE WANT TO COMPREHEND PHYSICAL REALITY, WE CANNOT CREATE A TOOL TO HELP US COMPREHEND THAT REALITY AND THEN ABDICATE COMPREHENSION TO THE TOOL.

One elementary particle defined non-existence, a second, uniformly spaced from the first, gave us the ability to measure distance.

The mistake we have made by believing that we could define the universe by mathematics is to take the measurement that exists between the two elementary particles and apply it outside the area between the two.

We are applying a system we created to measure and understand physical reality where there is no physical reality.

This, of course, is senseless, and when it gets senseless results, we call sense mere intuition, and the measurements reality.

Which, with the goal being comprehension, is both incomprehensible and reprehensible.

Once we have the two elementary particles, we can alter their position with respect to one another.

If we take one elementary particle and move it closer to the other elementary particle; we have diminished measurable space, physical reality, in the process re-expanding non-existence. But in doing so we have created time.

Time is the interval that occurred between the two positions of the elementary particles.

Time exists only in so far as there is movement that produces an interval between two events.

When the movement has stopped, there is nothing to measure.

The mistake we have made by believing that we could define the universe by mathematics is to take the measurement that exists because of the movement of the two elementary particles and apply it without reference to that movement. We are applying a system we created to measure and understand an event in physical reality where there is no event in physical reality.

This, of course, is senseless, and when it gets senseless results, we call sense mere intuition, and the measurements reality.

Which, with the goal being comprehension, is both incomprehensible and reprehensible.

Now, when we moved the elementary particle into a new position, we did so at a constant rate. We could have moved it in a manner that increased the interval, or decreased the interval.

When we created time by moving the distance between two elementary particles with respect to each other, we created a third measurement that is dependent on the first two, the rate of change between two events.

The formula that gives us our information about physical reality, time times rate equals distance is derived from the existence of that physical reality.

The mistake we have made is believing that we could define the universe by mathematics rather than by using mathematics to describe the existence and movement of matter that exists in that universe, and then use our reason to comprehend that existence and movement. We are applying a system we created to measure and understand physical reality as a substitute for the comprehension of that physical reality.

This, of course, is senseless, and when it gets senseless results, we call sense mere intuition, and the measurements reality.

Which, with the goal being comprehension, is reprehensible.

Now that we have our elementary particle's dimensions, it is only large enough to define non-existence, we need to define its properties, those characteristics that are going to allow it to become the universe we comprehend and the mind we use to comprehend it.

HISTORICALLY, THE PROPERTIES ASSIGNED TO ELEMENTARY PARTICLES REFLECT THE PARTICLES EXPLAINED. AS A SINGLE ELEMENTARY PARTICLE HAS TO REFLECT PHYSICAL REALITY, ITS PROPERTIES MUST BE DERIVED FROM PHYSICAL REALITY.

We are not going to derive those properties from non-existence, from some sort of derived precepts or universally accepted laws, assumptions or generalities.

We can't pull properties out of thin air and accept them because they are evident to all mankind!

We can define the size of the elementary particle in non-existence, because that size is governed by non-existence.

When it comes to properties, however, we have to look to reality.

If we want to describe reality, we have to determine what is universal about that reality, and then synthesize those universals into properties that could define the elementary particle.

Once we do that, we can then see if those properties will make the elementary particle behave in such a manner as to result in what we can measure, using our knowledge of time times rate equals distance, and two plus two equals four, the existence of matter and the changing relationships of matter that result in measurable physical phenomena.

This really isn't a very difficult thing to do.

The first thing that we know about matter is that it exists.

This may be a deceptively simple statement to make, but it is so obvious that it might escape notice.

THERE ARE NOT VERY MANY COMMONALITIES WITH RESPECT TO PHYSICAL REALITY.

External reality is a fact.
It is not something that we make up in our mind.

Our mind developed so that the external reality of our bodies could move safely through external reality. The mind constructs a picture of external reality so that it can direct the body that contains it how to move within that external reality to insure its survival.

Physical reality exists, and in so doing, it must be made up of not just one or two elementary particles, but uncountable elementary particles that are adhering together.

If a single elementary particle is only big enough to define non-existence, then the number it takes to make up the head of a pin must be beyond comprehension, even though the number must exist.

It is really not necessary to know how many of these elementary particles actually do make up the head of a pin, or of anything else.

What we need to know is that they do make up the head of a pin and everything else.

They are not moving around in space, shooting off in all directions, or doing anything else but making up the matter to which they give physical presence.

THE EXISTENCE OF MATTER PRESUPPOSES THAT SOMETHING IS HOLDING IT TOGETHER.

If that is the case, then there is some force that is causing them to adhere together so that there are enough of them in one place to make up the Earth we are standing on, the air we are breathing, the lungs that we are breathing with, and the mind we are using to observe ourselves doing all of this.

When we are trying to discover the property of an elementary particle, we don't want to start with the fact that some of them form into iron, others into gold, and still others into trees.

All of these things are marvelous, but diversity can conceal as well as reveal understanding.

The thing that all matter has in common is that it is held together.

It exists.

If we know that matter is held together, we know that something is holding it together.

Whatever is holding it together must be common to all matter.

There is no sense in creating one thing to hold a tree together and another to hold a stone together.

Whatever is holding everything together might as well be the same thing.

What could that be?

Gravity?

Gravity is now conceived to be a force that acts over a distance without existing within that distance.

This is incomprehensible, and we are trying to comprehend, not obscure our lack of comprehension.

How about magnetism?

Magnetism is born of concepts of positive and negative, plus and minus, north and south that make it incomprehensible. After all, what type of comprehension is it that has magnets made up of molecular magnets with the next level of understanding being atomic magnets.

No comprehension at all!

Another key turned in the ignition.

We are trying to comprehend, not obscure our lack of comprehension.

We could solve our problem of why matter clumps together simply by saying that every elementary particle has a property that makes it want to occupy the space of every other elementary particle.

PROPERTIES ARE CONCEPTS. THE CONCEPT THAT EVERY ELEMENTARY PARTICLE WANTS TO OCCUPY THE SPACE OF EVERY OTHER ELEMENTARY PARTICLE IS A CONCEPT THAT CAN BECOME THE PROPERTY OF AFFINITY PROPENSITY, THE PROPERTY OF THE ELEMENTARY PARTICLE THAT HOLDS IT TOGETHER TO FORM MATTER.

This would result in matter, and it would also conform to the concept of the elementary particle defining non-existence. After all, there is no reason to define non-existence in two places.

If one particle existed, and another came into existence, it would attempt to define non-existence as efficiently as the first elementary particle, and therefore it would try to occupy the same space as that elementary particle.

We don't have to speculate on why elementary particles come into existence to assign a property to them, and if we want to state that a property of the elementary particle is a propensity to occupy the space occupied by every other elementary particle in existence, then this propensity, we can call it the affinity propensity, is defined and becomes one of the two properties of the elementary particle we are going to use to build the universe and the mind that comprehends it.

With the glue that holds matter together, the affinity propensity, we have one half of the properties of the elementary particle.

The affinity propensity is the tendency of every elementary particle to occupy the space of every other elementary particle.

The property of affinity propensity is what makes elementary particles clump together to become the physical matter we exist in and are.

What is the other property of the elementary particle we are going to use to explain physical reality?

We synthesized the first property from looking at what matter is.

It is a group of elementary particles held together by their affinity propensity, the propensity to occupy the space of every other elementary particle.

Since it is matter the physical operation for which we are seeking a rational explanation, we should be able to synthesize the second of the two properties of the elementary particle by looking at matter.

Looking at what matter *is* allowed us to synthesize its property of affinity propensity.

We can find the second property of the elementary particle by looking at what matter *does*.

What does matter do?

We could say that matter moves.

But that is an inappropriate level of explanation.

Matter may well move, but it does so only because forces act upon it.

Matter moving is not something that matter itself does.

It doesn't move by itself.

So what really does matter do?

It combusts.

As a matter of factual congruity, there are only two things that are happening in our part of the universe, matter undergoing combustion and matter moving.

MATTER EXISTS. MATTER MOVES. MATTER COMBUSTS. OF THE THREE, ONLY MOVEMENT IS NOT INHERENT IN MATTER. EXISTENCE AND COMBUSTION CAN PROVIDE THE TWO PROPERTIES, AND ITS MOVEMENT CAN RESULT FROM ITS COMBUSTION.

Nothing else!

That alone would warrant examining the connection we made through much more measured detail in the last chapter, the fact that its movement matches the physical expansion of the electromagnetic spectrum.

But so what if matter combusts?

How does that provide us with the second property of the elementary particle?

The concept today of combustion is similar to putting a log in the fire to cook the tribal dinosaur. The thing burns down into charcoal, emitting light in the process.

The atoms and molecules that made up the log are around in the form of ashes, gasses and smoke.

The light has gone somewhere, probably the ends of the universe, so that it can curve back on itself in order to keep the fire stoked.

IF A SINGLE ELEMENTARY PARTICLE IS TO ACCOUNT FOR PHYSICAL PHENOMENA, THE NOTION THAT LIGHT IS MADE UP OF SOMETHING OTHER THAN WHAT THE MATTER THAT PRODUCES IT IS MADE UP OF WILL HAVE TO BE ABANDONED.

But one thing is certain in modern science.

While there is no present concept that can describe what light is, it isn't part of the matter that made up the log!

This is because, while we don't have the foggiest idea what light is, we do know that it is produced by an electron shifting orbit as it rotates around the nucleus of a conceptual atom designed to explain chemical reactions.

This, of course, is becoming incomprehensible, as usual, so we have a good idea someone is attempting to disguise their lack of comprehension.

Actually, there is no reason why what is emitted in the

combustion process is not what makes up the matter undergoing combustion.

To assume otherwise would be to place an unnecessary burden on the attempt to understand physical phenomena, much like assigning different elementary particles for different effects.

When matter combusts, it is emitting the elementary particles that make it up.

A fire in the fireplace might not be hot enough to break down more than some of the molecules that make up the log, but that doesn't mean that some of the elementary particles that formed the molecular structure weren't emitted in the process.

And the hotter the combustion process, hotter being another way of saying more rapid, the more rapid the combustion process, the greater the number of elementary particles that will be emitted is, and the further the matter will break down until you reach the point of fission where some of the actual matter disappears, converted directly into energy.

Energy?

Under today's incomprehensible concepts, energy is something other than the matter that produces it although one can be converted into the other by the famous conceptual equation e(energy) equals the amount of matter (m for the incomprehensible mass) times the speed of light squared (c^2, at one hundred and eighty-six thousand miles per second, another number like the amount of elementary particles it takes to make up the head of a pin).

What $E=mc^2$ is saying is that an awful small amount of matter can be converted into an awful lot of energy if all the matter is consumed.

Thus, if the elementary particles that make up the head of a pin are all of a sudden released in the combustion process, there is going to be a lot of light.

And that is what the combustion process produces.

Light!

COMBUSTION PRODUCES LIGHT, THE ELECTROMAGNETIC SPECTRUM, WHICH IS COMPOSED OF ELEMENTARY PARTICLES TRAVELING AT THE SPEED OF LIGHT.

Light is the electromagnetic spectrum, or at least the portion we can see, and combustion produces the expanding sphere that diminishes inversely with the distance from its source, the field in which matter moves at an accelerating rate that is also inverse with its distance.

But we are not yet at the stage of describing what it is about the expansion of light that would cause matter to be drawn back to the source of the combustion that is producing the light.

We are trying to find the second property of the elementary particle so that the two hypothetical properties can become a hypothetical congruity.

If all combustion produces light and light is made up of the elementary particle, what is it about light that can tell us something about the elementary particle?

The one thing we know about light is that it travels at a uniform speed, about one hundred and eighty-six thousand miles per second.

THE ELEMENTARY PARTICLE THAT MAKES UP LIGHT CAN SLOW DOWN AND SPEED BACK UP DEPENDING UPON THE MEDIUM IT IS TRAVELING THROUGH. A SINGLE PARTICLE IN SPACE HAS NO MEASURABLE MOTION. THE CONSTANT MOVEMENT OF AN ELEMENTARY PARTICLE WITH RESPECT TO MATTER CAN BE THE SECOND PROPERTY OF THAT PARTICLE.

If it passed through water, that speed is reduced.

The light slows down.

When it comes back out of the water, the light regains its speed.

When the elementary particles that make up light are restrained into the form of matter, they are not moving at all.

But when the combustion process occurs, they are released from being held into the matter and they instantly begin to travel at a hundred and eighty-six thousand miles a second.

This property of speed, in fact, begins to look as if it is inherent in the elementary particle.

The second property could therefore be the property of

speed.

Unless an elementary particle is otherwise restrained, it would always travel at a rate of a hundred and eighty-six thousand miles per second.

In fact, as far as the particle is concerned, even though it would be traveling at a hundred and eighty-six thousand miles per second with respect to elementary particles immobilized by their affinity propensities in the form of matter, that particle is at rest.

The second property of the elementary particle would then be the property of at-rest motion, the velocity of the speed of light while the elementary particle is at rest.

WE WILL USE AN ELEMENTARY PARTICLE WITH THE OPPOSING PROPERTIES OF AT-REST MOTION AND AFFINITY PROPENSITY TO CREATE A CONSISTENT PICTURE OF PHYSICAL REALITY.

These two properties, affinity propensity and at-rest motion, are diametrically opposed to each other.

If the property of affinity propensity dominates, matter would exist.

If the property of at-rest motion gains the upper hand, matter will dissipate, its elementary particles released by combustion, flying off in every direction.

Let's create a laboratory in non-existence and take a closer look at how elementary particles only big enough to define non-existence, each with an affinity propensity to occupy the space of every other elementary particle that is opposed by its at-rest motion of the speed of light, might form into the matter that exists in physical reality.

5. AFFINITY PROPENSITY: THE ATOM

In the beginning, there was light, or at least energy, as that is understood, or at least not understood.

Energy is a good place for the present conceptualists to start the universe because their concept of energy is the same superstitious mumbo jumbo they claim for primitive religions, just a little more sophisticated.

Consider gravity, the universal invisible energy: It's not something, but rather a property much like the color of gold. Yellow is a property of gold, gravity of matter, analogous to the key running a car.

> **THE CONCEPT OF GRAVITY EMBODIES ALL THAT IS ADMIRED IN THE OCCULT COMMUNITY, BUT THEN IT WAS CONCEIVED BY AN ALCHEMIST SOME THREE CENTURIES AGO.**

Next, while gravity governs the universe, it is never used up. This means that the same gravity that is causing the moon to fall to the Earth is making you stick to the Earth. By exerting its force, it does not diminish itself. This is a truly remarkable characteristic, sort of like a perpetual motion machine.

Gravity also has the ability to flow back and forth between masses. If you lift a rock, you're increasing its mass. Where is the proportional increase in gravity coming from? If something can't be used up in its use, then it doesn't have to come from somewhere in its increase, I guess.

Finally, and best of all, gravity doesn't exist between the matter that it effects.

I could add a few, such as the ability of gravity to increase with speed, and its equivalence with acceleration, but once you are totally confused, what reason is there to make total confusion a null confusion and increase it?

Modern science exists in a world of man made equations, and that world has never examined force at a level of understanding that is different from the level needed to know that a key runs a car.

So what better thing to begin the universe with, than light, energy, the electromagnetic spectrum, that which is convertible in the great conceptual equation $E=mc^2$ into the matter we see in the universe?

Thus, some eighteen billion years ago (a little longer in Venusian years, a little less in Martian ones) all of this energy that filled the universe (which was defined by its infiniteness) converted itself in one big bang into atoms of the simplest element possible, hydrogen, which immediately began to form into heavier elements.

The atoms that resulted, it was discovered some hundred years ago, had a neutron in their nucleus to explain mass, one or more electrons in one or more shells that explained chemical reactions, the reason the atom was conceptualized in the first place and a proton in the nucleus, which had to be there because the electron was negative, and wherever there was a negative in a structure, according to the law of magnets, there had to be a positive.

Unfortunately, for the universe, which started out with the simplest atom, the hydrogen atom, the only way to get to heavier atoms would be to combine the nuclei of the simple atoms to make heavier atoms.

The law of magnets ruled that opposites attract and likes repel.

How could nuclei with a positive proton combine with another nucleus with a positive proton when likes repelled.

What a problem!

CURRENT CONCEPTS REQUIRE SIMPLE MATTER TO FORM INTO MORE COMPLEX MATTER. NEWTON WOULD HAVE TURNED LEAD INTO GOLD. MODERN SCIENCE WOULD USE HYDROGEN IN PLACE OF THE LEAD.

But never worry. When there is no requirement for conceptual consistency, any explanation that can be quantified into an equation is acceptable.

Thus, when the density of the gas that was formed out of the energy after the big bang passed about one hundred times the density of water, why the proton/proton repulsion becomes a proton/proton attraction.

This happens because masses of gas, the well-established source of planetary formation, coalesce out of the energy converted hydrogen atoms, compresses them, heats them up, starts to burn them and then gets so hot that the transition to nuclear energy occurs.

But this is not planetary formation.

This is star formation!

First, the heavier elements have to be formed as the stars burn, and then, as they age, the hot interior heats the cooler shell where the heavier elements bubbling up in the interior are collecting, and this shell expands, hurling these heavier elements into space to become, as the poets say, the star dust that makes up humanity.

It's this dust that starts swirling into planetary nebulae, again fortunately differentiating itself so that the particles formed into the solid planets close to the sun and the gas formed into the gas planets not so close to the sun.

This hodgepodge of ideas, brought together over the last three hundred years, its only deference to consistency a slavish worship of the mass/gravity concept, certainly validates the major excuse in science that we all stand on the shoulders of our predecessors.

And if we all continue to stand on the shoulders of our predecessors, our minds will never evolve to comprehend the universe, and in failing to do so, will be useless to the evolution of the universe.

If the basic work of science is to come up with a consistent picture of physical phenomena, then it's hard to credit the uncritical acceptance of this gibberish as a good faith attempt.

In structure, it parallels the past. Before Newton came along, the bible told us that in the beginning there was light.

Newton, the alchemist, came along obsessed with turning base metals into gold.

Today's scientist is obsessed with finding formulae that would show how the hydrogen atoms are converted to things like lithium, boron and beryllium in an effort to get to the gold of carbon, the basis of life.

And as they relentlessly pursue symbolic textualism, they forget that testing ideas against observation only works for verifying measurable facts, or perhaps, believing that is the correct procedure, render the comprehensible incomprehensible by failing to test the ideas against consistency, not consistency with some occult idea about matter, but with each other.

The only way to obtain consistency is to require that the picture fits together. We can't have an atom doing one thing one place, another in another place. If a proton is positive, and positive means that it repels, then it doesn't all of a sudden attract when the conclusion requires it to attract. Forming solid planets out of solid matter, and gas planets our of gas is nothing more than turning the ignition to start the car.

And assigning a time for the beginning of the universe?

I don't think so.

We are attempting to achieve consistency by synthesizing an elementary particle from measurable reality and then seeing if we can construct that measurable reality out of the particle.

An area of total incomprehensibility is the plus/minus, positive/negative, north/south derived from the magnetic, electric, atomic, electromagnetic spectrum confusion.

The idea of opposites arose from the physical observation of the behavior of magnets. Magnetic material, if broken, will recombine with an attraction at the break, but will repel at the ends opposite the break.

This is all well and good, but when it was found that a difference in potential between two elements created electricity, the idea that something was being attracted from the atom in one place to the atom in another imprinted the idea of the magnet on the concept of electricity, and then because atoms made up the elements, the atoms themselves.

THE MYSTERIOUS BEHAVIOR OF MAGNETIC MATERIAL IMBUES SCIENCE WITH A MYSTICISM THAT COMPLEMENTS THE OCCULT NATURE OF THE CENTRAL MASS/GRAVITY CONCEPT.

Atoms don't require polarity, which is the designation given when elementary particles move from one place to another. Polarity results from such movement because, while at any point, the particles are moving both to and away, there has to be a designation when the particles are moving through matter.

Thus, the North pole of the Earth is defined by elementary particles moving through the center of the Earth from south to north, out the "top" of the planet, and, like a bar magnet, circling around and re-entering at the south pole.

The north pole of a magnet is defined by elementary particles moving out of that end, and re-entering at the south end.

It's fairly obvious why these flows of elementary particles would either align or oppose alignment depending on the physical positioning of the matter around which they are moving (making the north pole of the magnet actually point toward the north pole of the planet).

THE BEHAVIOR OF MAGNETS BECOMES DESCRIBABLE IF THE LINES OF FORCE ARE ELEMENTARY PARTICLES ORBITING THE MAGNETIC MATTER AT THEIR AT-REST SPEED.

An atom, however, is a self contained unit. It is a structure made up of elementary particles, and elementary particles orbit it in clouds resulting in the atom's ability to engage in chemical reactions.

But it does not have, nor does it need, polarity.

Thus, the atom doesn't need a proton to balance the electron!

The orbiting elementary particle stays in place nicely

by its at-rest motion at the speed of light balancing its affinity propensity to occupy the position of each of the elementary particles that make up the nucleus of the atom.

Because gravity is not a property of mass, or at least, there is no reason to assume that it is, there is no reason to have a neutron in the nucleus of the atom that is going to form in non-existence from elementary particles as a result of their affinity propensities overcoming their at-rest motion.

THE OPPOSING PROPERTIES OF AT-REST MOTION AND AFFINITY PROPENSITY ELIMINATE THE NEED FOR THE CONFUSING CONCEPTS THAT GREW UP FROM OBSERVING THE BEHAVIOR OF MAGNETS.

We need neither the proton nor the neutron, hallmarks of the classic atom, to make up the atom.

They simply aren't required for the atom to describe physical reality!

To understand how atoms would form, it is necessary to look briefly at the peculiar nature of the mass/gravity concept where gravity is produced without being used up.

Why, in a universe where there is never anything used without it being used up, someone would come up with such a concept shows the creativity of the human mind unhindered by factual reality.

If we focus for a second on gravity's ability to attract infinite matter without expending energy, we will realize that the opposite is true, that when energy is expended in holding something together, then it is expended, used up.

With the elementary particle, if one comes into existence in non-existence, we would have no way of knowing what its properties were.

As far as its at-rest motion at the speed of light is concerned, it might as well be still. Without a reference point for motion, there is no motion.

As far as the affinity propensity is concerned, there would be no way of knowing whether it existed because there would be no other elementary particle for it to attempt to occupy the other's space.

AN OCCULT CONCEPT CAN HAVE OCCULT PROPERTIES. THEREFORE, GRAVITY HAS THE PROPERTY OF NEVER BEING EXPENDED.

When we create a second elementary particle in our non-existence, which is now being defined by the elementary particles that occupy it, then the two particles are going to seek to occupy the same portion of that non-existence.

At the same time, however, the at-rest motion of each will attempt to oppose that combination that would result if they succeeded in occupying the same space.

If the two elementary particles came into existence with the same motion, side by side, then they could combine without too much opposition from their respective at-rest motions.

If twenty came into existence, and combined so that the at-rest motion of each had been overcome by the affinity propensity, then the amount of work it would take on the part of the affinity propensity of each to keep the other elementary particles from flying off in twenty different directions would be greater than it would be if the twenty were moving in the same direction.

The point is that the two properties we have induced for the elementary particles oppose each other. The affinity propensity can overcome the at-rest motion, and in the process, form into physical matter, but only until the affinity propensity is used up after which it can no longer cause elementary particles to become part of the structure.

Because each elementary particle has the same affinity propensity, the structures that are able to form by overcoming the at-rest motion will be identical, composed of the same number of elementary particles.

The structure that results can be thought of simply as a unit composed of stationary elementary particles formed into units surrounded by a cloud of the elementary particles that make up the units: unit, units, cloud!

The atom can be thought of as a nucleus of units that has a cloud of orbiting elementary particles moving around the nucleus.

The concept of matter forming in the primordial explosion into the simplest element and then somehow proceeding to the complex comes from the fact that the surface of the sun is populated by units of elementary particles in their last stages of being broken down and emitted in the form of the electromagnetic spectrum.

That occurs in the highest field possible.

In the absence of a field, in the non-existence where our matter is forming, the elementary particles will overcome their at-rest motion in the attempt to occupy each others' space.

As the first two come together, the resulting affinity propensity is twice as much as the affinity propensity of each separate particle, but a portion of that affinity propensity is used up in holding the structure together, in preventing the two elementary particles from regaining their at-rest motion.

THE SURFACE OF THE SUN APPEARS TO BE HYDROGEN BECAUSE THE INTENSE FIELD OF COMBUSTION HAS REDUCED COMPLEX ATOMS DOWN TO THEIR INDIVIDUAL UNITS.

The affinity propensity, however, of the structure is greater than the at-rest motion of any individual elementary particle, and the elementary particles race to the structure to occupy its space.

Each adds to the structure, but diminishes the overall affinity propensity of the resulting structure.

Soon, when heaven knows how many elementary particles clump together into the structure, the structure has used up all of its affinity propensities just holding itself together.

We don't know how many elementary particles it takes to make the resulting structure, but we do know that it is always precisely the same because each elementary particle has the identical amount of affinity propensity.

If we had an inexhaustible supply of these elementary particles coming into existence, then they would form into units containing an identical number of elementary particles.

But, while the completed units will no longer have enough excess affinity propensity to be able to cause additional elementary particles to overcome their at-rest motion, they will still have an excess of affinity propensity.

UNITS OF ELEMENTARY PARTICLES ARE IDENTICAL BECAUSE THEIR SIZE IS LIMITED BY THE SUM OF THE INDIVIDUAL AFFINITY PROPENSITIES REQUIRED TO HOLD THEM TOGETHER.

This excess of affinity propensity will cause the units themselves to clump together into a nucleus until the excess affinity propensity of the units can no longer cause another unit to adhere to it. Because the at-rest motion of the elementary particles in the units has already been overcome, the number of units that would come together would be determined by the excess affinity propensity of each of the units.

As each of the units would be composed of the identical number of elementary particles, and thus have an identical amount of excess affinity propensity, the resulting nuclei would all be identical.

But even though the nuclei could no longer attract and hold an additional unit, it would still have an excess affinity propensity.

That excess affinity propensity would not be sufficient to overcome the at-rest motion of elementary particles, but it would be sufficient to partially overcome it, and the nuclei would attract a cloud of orbiting elementary particles, orbiting the nuclei, sharing orbits with the units that make up the nuclei.

MATTER HAS TO FORM SOME WAY BECAUSE IT EXISTS. WITH ELEMENTARY PARTICLES FORMING UNITS AS A RESULT OF AFFINITY PROPENSITY AND THE UNITS FORMING INTO NUCLEI WITH CLOUDS OF ORBITING ELEMENTARY PARTICLES, MATTER FORMS INTO THE MOST COMPLEX ATOM ALLOWED BY THE EXCESS AFFINITY PROPENSITY.

Thus, in our area of non-existence into which were pouring elementary particles that had the opposing properties of at-rest motion and affinity propensity, atoms with an equal number of identical units in the nucleus would be forming with clouds of orbiting elementary particles surrounding them.

Unit, units, cloud.

The presence of the clouds, however, would also create affinity propensities, and as the clouds were contiguous, the nuclei would start to share elementary particles in the outer areas of the clouds. This would go on until the affinity propensities of the orbiting elementary particles had normalized the distances between the nuclei.

The atoms would form into matter.

What type of matter?

The most complex atom that could exist in non-existence, as determined by the elementary particle's at-rest motion and affinity propensity.

EVEN ORBITING ELEMENTARY PARTICLES HAVE EXCESS AFFINITY PROPENSITIES WHICH CAUSE THE RESULTING ATOMS TO COMBINE TOGETHER INTO PHYSICAL MATTER.

We could imagine eons of experimentation, first with an at-rest motion so great that only miniature atoms would form, next with at-rest motion so little that only giant atoms would form, first one way, then the next, until just the right balance was reached to form the universe we perceive and make up (and which might not be the right mix at all, perhaps just another failed experiment).

The atoms of matter that form, form with the highest number of units in their nuclei so that they break down in the combustion process into simpler atoms.

This is the opposite of the current concept of matter

forming at its simplest and combining in the combustion process to make elements with a greater number of units in the nucleus.

But then, the concepts of modern science do seem to mirror reality, with the image being exactly backwards.

As the atoms formed and combined, they would grow into massive structures that could reach the size of a star.

When the star finally ignited, the atoms would begin to break down into their component units on the surface of the combustion.

MODERN SCIENCE WOULD HAVE STARS BURNING FROM THE INSIDE OUT BECAUSE OF THE THEORETICAL PRESSURE CAUSED BY THE WEIGHT OF ALL THE LAYERS OF MATTER PRESSING ONTO THE CENTER. THE CONTRARY IS WHAT WE SEE, SURFACE MATTER BREAKING DOWN IN THE COMBUSTION PROCESS.

And the combustion would be on the surface, not as present concepts put it, from the center out.

Again you have to shake your head in wonder at a science that creates something that burns from the inside out when in reality there is nothing that burns from the inside out. The reason modern science has it burning from the inside out is that the hypothetical gravitational forces create such intense pressure at the center of the sphere that the pressure evidences itself in heat.

However, it is well understood that were mass/gravity

a reality, in a spherical body the pressure would not be at the center of the matter, but rather at every point on the surface, the center being a point at which all gravitational attraction canceled out and thus created weightlessness.

Where reality is blinded by mathematics, the absurd becomes commonplace.

THE EARTH IS COMBUSTING FROM THE OUTSIDE IN, WITH THE CRUST COOLING OVER THE MOLTEN ROCK WHICH RESTS ON A LAYER OF ACTIVE COMBUSTION WHICH SURROUNDS A STABLE NON-COMBUSTING CENTER OF HIGHLY COMPLEX ATOMS. MODERN SCIENCE CONSIDERS THE EARTH TO HAVE MOLTEN LAYERS WHICH DO NOT UNDERGO THE COOLING PROCESS AS IT MOVES THROUGH ABSOLUTE ZERO DEGREE SPACE.

The structure formed out of these complex atoms would continue to grow until its equilibrium was disturbed. That disturbance results in ignition and combustion.

The size of the resulting combusting matter could be as big as a star such as the sun, or the size of a planet, such as the Earth, where the combustion has occurred from the surface inward, the surface congealing as it cooled.

What happens to the most complex of atoms possible as they undergo the process of combustion?

On the surface of the combustion, the atoms would break down into the individual units. The units would measure out at the one or two units we associate with hydrogen and helium, and thus, the concept builds up that the sun is made out of hydrogen, it burns by turning hydrogen into helium, and ninety-nine percent of the universe is composed of these two light elements, all other elements being formed by their combining together to make elements with more units in the nuclei.

Looking at reality, the surface of the Earth, we can see that just the opposite occurred.

As the most complex atoms began to undergo combustion, they began to break down into more simple elements in a predictable geometric pattern. If the most complex atom has two hundred units in the nucleus, then it might break down into halves, and thirds, with the halves and thirds recombining with fourths and fifths so that there would be an orderly progression of elements with units in the nuclei spaced from one to two hundred, or less, depending on the resulting rate of combustion.

This process has its analogy with chemical reactions on Earth, where various elements combine and recombine with other elements under conditions controlled by combustion.

If we shortcut the process of the formation of the atom with the highest possible number of units in its nucleus to the break down of those atoms in the process of combustion into atoms with fewer units in the nucleus, we can scoop out a hydrogen atom and a helium atom and take them back into non-existence with us to gain an understanding of the mass/momentum morass the concept of mass/gravity has gotten us into.

If we put these atoms side by side in non-existence, we can see one atom made up of one unit with a cloud of elementary particles surrounding it, and another made up with two units held together by their affinity propensity, with a cloud of elementary particles also surrounding it.

Normally, the existence of these atoms would be dependent on the combustion field in which they exist, the subject of Chapter Nine, but for now we are going to allow them to exist in non-existence side by side.

THE PERIODIC TABLE OF ELEMENTS REFLECTS THE ORDERLY BREAKDOWN OF COMPLEX ATOMS IN A STRONG EMISSION FIELD RATHER THAN THE HAPHAZARD BUILD UP OF COMPLEX MATTER EJECTED FROM THE CENTER OF AGING STARS.

What do we know about them?

We know that if they exist the way they do, then the affinity propensities in each balance the at-rest motion of the elementary particles that make them up.

The units are conglomerations of elementary particles that have had their affinity propensity overcome their at-rest motion.

This is true of all three units, the one that makes up the hydrogen atom and the two that make up the helium atom.

We know that the unit in the hydrogen atom has the excess affinity propensity balanced by orbiting elementary particles.

We know that the helium atom has two units. Those two units first use up affinity propensity to hold themselves together. Once they are held together, then the excess affinity propensity of the two together is balanced by orbiting elementary particles.

The two atoms are in perfect equilibrium, the equilibrium being between the force of the affinity propensities of the elementary particles that holds the structure together, and the force of the at-rest motion that would, but for the affinity propensities, allow all the elementary particles to fly away in a flash of combustion.

THE FACT THAT ATOMS WITH DIFFERENT ATOMIC WEIGHTS MOVE FREELY IN THE FIELD OF ATTRACTION IN THE SAME MANNER, BUT MOVE AGAINST THAT FIELD DIFFERENTLY RESULTS FROM THE FORCE HOLDING THE UNITS OF THE ATOM TOGETHER. IT PROVIDES A CLEAR WINDOW INTO THE NATURE OF THE ATTRACTIVE FORCE.

Do we have weight?

No weight!

Each atom has the forces affecting it in perfect equilibrium, they are current forces, there is nothing affecting that equilibrium.

There is no weight.

Now, let's assume an attractive force disturbs the atoms.

The way the attractive force affects the atoms is that it makes them move.

If the attractive force acted on the individual units at the center of the cloud of orbiting elementary particles, then the two atoms would move toward the source of the attractive force in an identical manner.

SCIENCE CONSTRUCTS ITS CONCEPT OF THE UNIVERSE ON THE BASIS OF MEASURING THE MOVEMENT OF MATTER IN THE UNIVERSE WITHOUT EXAMINING THE CURRENT FORCES THAT CAUSE THAT MOVEMENT.

This is the analogy to dropping a lead ball and a feather. Both move toward the source of attraction in the same manner because the force is acting on the individual units.

However, what happens if we attempt to stop the movement once it has started?

If we try to stop the hydrogen atom, we are attempting to stop a single unit as it moves with the force.

However, if we attempt to stop the helium atom, we are attempting to stop two single units, but those units are tied together by the affinity propensities of the respective units.

Thus, it is going to take twice the amount of effort to stop the helium atom than it is going to take to stop the hydrogen atom.

If, instead of attempting to stop the motion of the two atoms, we simply put a surface between them and the source of attraction, then they would come to rest on that surface.

That is the position we are in on the Earth.

We start off moving things against the surface that exists between us and the source of the attraction, the combustion process that is occurring about a hundred miles beneath our feet.

BECAUSE SCIENCE FOCUSES ON THE MOVEMENT OF MATTER RATHER THAN THE FORCE THAT IS MOVING THE MATTER, IT INCORPORATES THE FORCE INTO THE MATTER ITSELF, EXPRESSING IT IN TERMS OF WEIGHT, MASS, INERTIA, MOMENTUM, POTENTIAL AND KINETIC.

We can't see the source of attraction, we just know that we always have to operate against it.

Instead of seeing atoms as balanced structures that exist without weight, we assign the force we constantly have to move them against to the atoms themselves.

Because the units in the atoms are bound together, the weight becomes the relative force it takes to move the different atoms against the force.

Instead of focusing on the force, and how it operates at the atomic level, we say that elements have weight, with the ones that are harder to move against the force having more weight than the ones that are not so hard to move against the force.

We define the matter by its movement within the field of attraction, rather than by the matter itself.

As soon as we assign weights that exist only as a reflection of the field of attraction, then we create qualities associated with the movement of that matter within the field of attraction, momentum, mass, equating the weight produced by acceleration with the weight created by the matter pressing on the surface.

In short, we start our analysis of physical phenomena with matter that is moving, or attempting to move.

We limit our focus to this narrow range of effect rather than cause, in the process distorting the picture we obtain of physical reality.

The focus puts the forces of movement as a quality of the matter, obscuring the fact that the forces involved merely move the matter. The more units there are bound together in the nucleus, the more difficult it is to move them in unison against a force, not because of weight, but, because of their internal bindings, all units have to be moved simultaneously against the force.

Newton's Second Law of impressed force works well under these conditions. Weight is known with respect to the existing field of attraction.

However, with respect to planetary motion, Newton's First Law of motion, that an object will move in a straight line unless a force acts upon it, has to be used together with the inverse square law to determine mass for the purposes of applying the Second Law of impressed force.

In other words, the success of the Second Law is used to gloss over the fact that the First Law does not apply in reality.

The question is not Galileo's "what makes an object stop", but Aristotle's "what makes it keep going".

Weight is the result of existing forces, and matter comes to rest with respect to existing, rather than historical forces.

Weight, to the extent it exists, exists as a result of the forces binding the units of the nucleus together rather than as a starting point for an analysis of physical phenomena.

MODERN SCIENCE HAS PROGRESSED ONLY TO THE POINT WHERE IT IS ABLE TO QUANTIFY THE MOVEMENT OF MATTER ON THE SURFACE OF A PLANE ON WHICH MATTER MUST BE MOVED AGAINST AN INVISIBLE FORCE.

The starting point for the analysis of physical phenomena is the atom in equilibrium.

Its movement comes later.

And then only in response to forces that exist at the time it is moving.

6. AT-REST MOTION: PRESENCE

We have seen that an elementary particle coming into existence in non-existence creates no problem of motion.

The motion of a single particle in non-existence has no reference point.

Without a reference point, there is no basis for determining motion.

When the elementary particles form together into a unit of matter, they are stationary with respect to one another. The property that makes them tend to occupy the same space has overcome whatever motion they normally have.

It is only when we have this structure, which may or may not be moving with respect to other forces, that we have a method of defining the motion of the elementary particle.

MODERN SCIENCE CONSIDERS ENERGY TO BE THE RESULT OF A HYPOTHETICAL NEGATIVE CHARGE CHANGING A HYPOTHETICAL ORBIT AROUND A HYPOTHETICAL NUCLEUS OF A HYPOTHETICAL ATOM IN A MANNER THAT EMITS A HYPOTHETICAL QUANTA WHICH IS A HYPOTHETICAL COMBINATION OF A HYPOTHETICAL PARTICLE BEHAVING LIKE A HYPOTHETICAL WAVE.

Considered in this light, at-rest motion is not that difficult of a concept to understand.

The reason that a picture of energy has eluded physical description is that we exist with gravity. In a field of gravity, everything we deal with has weight. Because both weight and the objects it is associated with are real, we tend to think of both in the same terms, with the matter itself being the cause of the weight, the mass/gravity concept.

Because everything we deal with has weight, if we want to move something, we have to overcome that weight.

For matter to fly through the air, or space, or anywhere, it has to be shot out like a bullet.

THE COMBUSTION PROCESS RELEASES ELEMENTARY PARTICLES SO THAT THEY MERELY REGAIN THEIR AT-REST MOTION, THE SPEED OF LIGHT.

Because there is no apparent analogy to light being ejected from the head of a match when we strike it, we conclude that light is something other than matter.

If it doesn't need to be ejected like a bullet from the end of a gun, or a rocket on a column of fire, then it is energy, not matter.

We have to deal with it the same way as we deal with matter, measure it, but other than that, it just exists.

This in spite of the fact that we know that energy is light, the electromagnetic spectrum, and although we don't know what that is either, we assign an elementary particle to it, the photon, and because we can measure the speed of light, we know that that photon must travel at that speed.

We also know that the light slows down when it enters a medium, and what is more important, regains its motion once it leaves the medium.

We also know that it bounces off things, just like billiard balls, but that it does not lose any of its speed when it does so.

Everything we know about the elementary particle that makes up light leads us to the conclusion that it moves in its normal state, the state at which it is in balance, at which it is expending no work, at one hundred and eighty-six thousand miles a second.

If its normal state is its measurable state, then it isn't a big jump to conclude that when it is used to make up matter, then its motion is restrained.

When its motion is restrained, the matter it makes up has a lot of pent-up energy.

When we strike a match, the sulfur at the tip is not ejecting the elementary particle that makes up light.

It is releasing it!

Combustion, then, the production of energy, becomes the process of releasing elementary particles so that they can once again regain their at-rest motion of one hundred and eighty-six thousand miles a second.

We have already seen that, as these particles are only large enough to define non-existence, the number of these particles that exist on the head of a pin, or in our case, the head of a match, make up a number that is so large it has no meaning to human comprehension.

But we do know that, whatever the specific process of combustion it is that is causing the particles to be released from the matter, to be emitted from the surface of the burning sulfur, the surface is undergoing combustion at such a rapid rate that there are more elementary particles on the surface that have regained their at-rest motion and are ready to be released than can be released.

The surface of the match head is a sphere with the elementary particles all attempting to move away from the center of the sphere at their at-rest speed, the speed of light.

ENERGY IS THE STRUCTURED RELEASE OF ELEMENTARY PARTICLES IN THE FORM OF THE ELECTROMAGNETIC SPECTRUM THROUGH THE PROCESS OF COMBUSTION.

Because the surface of the sphere is limited by its area, only the number of elementary particles that that area can accommodate at any one instance can move away from the surface of the combusting sulfur.

This means that, instead of elementary particles moving away from the match head in a willy-nilly manner, they have to wait their turn, so to speak, line up and wait until the ones in front of them have left the surface.

This process means that the release of the elementary particles will occur in an orderly fashion. The process results at any instance in the release of a sphere of elementary particles that moves away from the match head, expanding over the surface of the expanding sphere of light, the light diminishing inversely with the square of its distance from the match head, or for that matter the Earth, or the sun.

We will look closely how this orderly process occurs, and what it is about the process that causes the units in the nuclei of atoms to move in a direction opposite to the expansion process, in a later chapter.

For the purposes of looking at at-rest motion, and the presence it creates when it produces orderly flows of elementary particles, the presence that in turn results in field replacement, the process that we measure as combustion in its visible manifestation, we have to concentrate on the individual elementary particles that are emitted from the surface of the match.

COMBUSTION IS THE PROCESS OF FIELD REPLACEMENT, A PROCESS WHERE ONE SET OF AFFINITY PROPENSITIES REPLACES ANOTHER SET OF AFFINITY PROPENSITIES WITH LESS PRESENCE THAN THE FIRST SET.

The progression is: Flows, presence, field replacement.

At any point on the surface of the burning match, and at any instance while it is burning, an elementary particle is leaving the surface.

As it leaves the surface, it is following right behind the elementary particle that left the surface the instant before and is ahead of the elementary particle that will leave the surface an instant later.

If the sun is ninety-three million miles away from us, then the elementary particle that enters our skin as we sit on the beach is a part of an unbroken stream of elementary particles that stretch from the surface of our skin to the surface of the sun.

Every second that we sit in the sun, the infinitesimal section of skin that that particular stream of elementary particles is passing through is the host of a stream that is one hundred and eighty-six thousand miles long.

If we could determine how small something would have to be to be large enough to define non-existence, we could figure out how many elementary particles pass through the surface of the skin each second at each point, but that is the type of intellectual non-activity that has focused attention away from the conceptual understanding of what is actually happening at any point at which we can see the sun's light.

At any point in the sun's expanding sphere of electromagnetic radiation, including the point on our skin, a stream of elementary particles one hundred and eighty-six thousand miles long is passing by.

The number of elementary particles is, like the number on the head of the pin, uncountable.

It's just a lot of elementary particles.

But it's not the number of particles, which is incomprehensible, that we want to concern ourselves with.

It's what's happening at the point at which they are passing that is of utmost importance in our journey to constructing a consistent picture off physical phenomena.

What is happening at any point in a flow of elementary particles?

To understand what is happening at any point on a flow of elementary particles, we can transport the flow into our hypothetical non-existence the same way we transported the hydrogen and helium atoms so we could isolate and observe how they behave.

Isolating a process allows us to obtain a clearer picture of what might be going on.

EVERY SECOND WE ARE IN THE SUN, A STREAM OF ELEMENTARY PARTICLES ONE HUNDRED AND EIGHTY SIX THOUSAND MILES LONG IS REACHING EACH AREA THE SIZE OF AN ELEMENTARY PARTICLE ON THE SURFACE OF THE SKIN, THE UNCOUNTABLE INDIVIDUAL ELEMENTARY PARTICLES PASSING THROUGH, BEING DEFLECTED OR BOUNCING OFF DEPENDING ON WHAT IS OCCUPYING THAT SPACE.

If we isolate the flow of elementary particles in non-existence, the first thing we can say about it is that it is a single flow of elementary particles. It may, when it was closer to the sun, have been bundled up with other flows of elementary particles, but by the time we are analyzing it, we have isolated it and traced it back to its origin as a series of individual elementary particles released from the surface of the sun during the combustion process going on at the surface of the sun.

The stream of elementary particles now stretches out before us, from left to right, one hundred and eighty-six thousand miles of elementary particles passing the point we have identified each and every second.

What can we tell about that point?

That there is, at all times, an elementary particle occupying it!

As a matter of fact, this is the case with respect to any point on the entire stream of elementary particles.

PRESENCE IS THE EXISTENCE AT EVERY POINT IN THE FLOW AT EVERY MOMENT OF A PORTION OF THE FLOW.

At any point on the stream, there is always present a single elementary particle.

This is because as one elementary particle moves out of the space that makes up the point, another elementary particle immediately moves into the space.

As long as the stream of elementary particles is flowing past that point, there will always be an elementary particle at that point, and at any point in the flow of elementary particles.

It's not the same elementary particle.

But it is an elementary particle.

Because there is always an elementary particle at any point in a flow of elementary particles, each point on the flow has a presence.

If we were talking about the expanding sphere of electromagnetic emissions that surround the match head, or the Earth or sun, then we could say that at any point in that expanding sphere of electromagnetic emissions there is a presence, because at any point there is always a flow of elementary particles, and the elementary particles at any point are always leaving that point and being replaced by identical elementary particles.

IF THERE IS ALWAYS AN ELEMENTARY PARTICLE AT ANY POINT IN THE FLOW, THEN THERE IS ALWAYS AN AFFINITY PROPENSITY AT ANY POINT IN THE FLOW.

And we could also determine what the presence was with respect to any point in the expanding sphere relative to any other point in the expanding sphere because we know that the light, or the flows that pass any defined point diminish inversely with the square of the point's distance from the source of the light.

But what do we mean by presence?

What is the effect of presence?

Let's go back to our hypothetical stream in non-existence, and refine it a bit.

It is difficult working with something that diminishes

inversely with the square of its distance of travel. However, we want to change the presence of our hypothetical flow of elementary particles from one to two to four and so forth.

What can we use as a substitute?

We can use electricity!

Remember that we are using a single elementary particle to explain physical phenomena, so that what is good for light should also be good for electricity.

If we don't have a flow of elementary particles being emitted from combusting matter, we will have to find some other way to order them into a flow with presence.

We know that nuclei of atoms have an excess of affinity propensity that causes them to attract a cloud of orbiting elementary particles.

What would happen if we took a bunch of atoms and depleted them of their orbiting elementary particles?

We would have atoms with an excess of affinity propensity, the excess being affinity propensity that is not balanced by the at-rest motion of other elementary particles and which therefore would cause other elementary particles to want to occupy the space that the atoms occupied.

If we then connected the atoms with excess affinity propensity with a path to a pool of elementary particles that are being used to balance the affinity propensities in other atoms, some of those elementary particles might move down the path to the atoms with the excess affinity propensity.

I say might, because whether they would or not would depend on whether the excess affinity propensity was greater than the affinity propensities that were holding the elementary particles in their orbital clouds around the other atoms.

This difference is what we measure as a difference in potential, with the potential being the potential of the excess affinity propensity of the first group of atoms of attracting elementary particles out of the orbital clouds that make up the atoms with less of an excess affinity propensity, or no excess affinity propensity at all, when there is a path provided for them to move from one group of atoms to the other.

DIFFERENCES IN POTENTIAL IN MATTER ARE REALLY DIFFERENCES IN EXCESS AFFINITY PROPENSITIES WHERE MATTER WITH A GREATER EXCESS AFFINITY PROPENSITY WILL ATTRACT ELEMENTARY PARTICLES FROM MATTER WITH LESS OF AN EXCESS AFFINITY PROPENSITY.

What we have constructed is the beginning of a battery.

With a path, elementary particles begin to flow from atoms with less excess affinity propensity to atoms with more excess affinity propensity.

So now in our blackboard of non-existence, we have replaced the imaginary flow we transported from combusting matter with an identifiable flow connected to two groups of atoms with different potentials.

We have a flow of electricity.

For our purposes, it is going to be a flow of single electrons, moving at the speed of light, one hundred and eighty-six thousand miles a second.

At any point in the flow there is a presence of a single elementary particle because as one particle leaves the point, another enters it.

We know that the elementary particles are traveling at their at-rest motion, the speed of light.

What is going on with the elementary particles' affinity propensity, their desire to occupy the space of every other elementary particle?

The affinity propensity was being used to balance the atoms that occupied the space where the elementary particles were coming from.

They are moving toward the excess affinity propensities that exist in the atoms to which they are moving.

They are not using any of their affinity propensity to move because their at-rest motion at the speed of light is doing that for them.

So what is going on with the elementary particle's need

to occupy the space of every other elementary particle?

We can assume that some of it is used up in holding the stream together.

The particle is trying to occupy the space of the particle that is moving out of the space it is moving into, and vice versa, so that it has the same relationship with the particle that is entering the space it is leaving.

That certainly doesn't use up all the affinity propensities of the elementary particles that make up the flow.

INDUCTANCE IS FORMED BY ELEMENTARY PARTICLES ORBITING A FLOW OF ELEMENTARY PARTICLES.

If elementary particles are immobilized into a unit in the nucleus of an atom, then the resulting unit, being a stationary structure of elementary particles will attract other elementary particles into orbits around it until the combined total of the orbits creates an orbiting cloud of elementary particles.

Because at any point in the flow of elementary particles, there is always an elementary particle, we can treat that particle as if it were stationary.

It always has a presence at that point.

Thus, any point in the flow can be considered to be a stationary point of excess affinity propensity that would attract elementary particles into orbit around that point.

There is only one problem.

With respect to a stationary nucleus, the elementary particles can orbit in any which way they please.

With the case of the points on a flow, the elementary particles can only orbit at an angle to the flow, and because there is excess affinity propensity at any point in the flow, that angle can only be at right angles to the flow at any point in the flow because otherwise they would interfere with the elementary particles in orbits at the other points in the flow.

INDUCTION IS A SECONDARY FLOW OF ELEMENTARY PARTICLES THAT BALANCES THE PRIMARY FLOW OF ELEMENTARY PARTICLES.

In short, in a flow made up of a flow of single elementary particles, there would be an orbiting elementary particle at each point in the flow that there is a presence, an elementary particle continuously moving in and out of the point, and that elementary particle would be orbiting the point at a right angle.

This is the right hand rule of induction for a moving charge that is the explanation for electricity.

If we take the thumb of the right hand and point it in the direction of the flow, then the fingers curve around in the direction of the inductive, or secondary flow.

Why is this secondary flow moving at right angles to the direction of flow instead of left angles?

If we take our thumb and point it in the direction of the north pole of a rotating body in space, we find the rotational motion of the planets to be in the direction of the fingers also. It is easy to see why, once the rotation of bodies is established with respect to each other, they would resolve themselves into orbits around each other that followed the same rule, but why the rule in the first place.

Why doesn't the inductive flow move according to a left hand rule?

Why don't the planets rotate in the opposite direction?

This is a situation in physical phenomena where there is no symmetry.

This is a rare occurrence in a universe where there is no direction.

There must be something else about what is going on in the universe that also lacks symmetry.

While we are not going to delve that deeply into the physical cause of the right hand rule, I should say in passing that the obvious dissymmetry in the universe is that after matter ignites, it combusts, and the combustion is a process of cooling.

Because an object does not cool in a straight line, but rather oscillates between hot and cold, the elementary particles emitted into the cooling leg of the combustion cycle would exceed the number emitted into the heating leg simply because each cooling leg is one tick longer than the corresponding heating leg (the definition of cooling).

I bring this up because the right hand rule does not emerge directly from the existence of the two properties of the elementary particle, its at-rest motion and its affinity propensity.

What does emerge directly from the flow, however, is the inductive flow as the elementary particles orbiting the flow balance the excess affinity propensities in the flow because at any point in the flow there is a presence that translates itself into excess affinity propensity.

THE EXCESS AFFINITY PROPENSITY OF ANY FLOW IS PROPORTIONAL TO THE NUMBER OF ELEMENTARY PARTICLES THAT MAKE UP THE FLOW.

Because we are dealing with electricity, a man made flow of elementary particles, we can regulate it with precision.

We can go over to our rheostat and turn it up a notch so that instead of a single flow of elementary particles, we have a double flow.

At each point in the flow there are two elementary particles entering and leaving two points in the flow bound together by their affinity propensity, moving at their at-rest motion, and with a presence at any point in the flow that causes elementary particles to orbit that presence.

What happens with respect to the orbiting elementary particles that make up the secondary flows?

If the presence doubles, then the secondary flow doubles.

What happens if we double the flow once again, creating four flows with four times the presence?

When we increase the presence we increase the secondary flows to balance the excess affinity propensities.

The result is a secondary flow four times the original secondary flow.

The secondary flows, the inductance, is proportional to the presence in the primary flow, the moving charge in current conceptualization.

AS THE NUMBER OF ELEMENTARY PARTICLES IN THE SECONDARY FLOW DEPENDS ON THE EXCESS AFFINITY PROPENSITY IN THE PRIMARY FLOW, THE SECONDARY FLOW IS PROPORTIONAL TO THE PRIMARY FLOW.

We have electricity, but we have more.

We have an explanation for electricity that is the direct result of the interaction of elementary particles with the two properties of at-rest motion and affinity propensity.

But what is more, we also have seen that wherever there is a flow of elementary particles with these two properties, there will be a presence.

A presence that contains its own affinity propensities.

FIELD REPLACEMENT:

7. FIELD REPLACEMENT: COMBUSTION

We could continue to use our battery without a circuit, but for purposes of understanding field replacement, the process that results in combustion, we should put the process on a laboratory bench and create a flow of elementary particles in a conductor with a generator rather than a wireless battery.

The generator is a device that moves a wire in a secondary flow, the inductive flow of elementary particles that orbit the primary flow in the conductor at right angles to balance the excess affinity propensities in the elementary particles that make up the primary flow.

As the wire, which is a conductor, moves through the secondary flow, elementary particles are captured out of the secondary flow into the conductor.

> **FIELDS CREATED BY AFFINITY PROPENSITIES EXIST IN A HIERARCHY IN THE ENVIRONMENT WITH ELEMENTARY PARTICLES MOVING FROM AREAS OF LESS EXCESS AFFINITY PROPENSITY TO AREAS OF GREATER EXCESS AFFINITY PROPENSITY.**

This capture is not the result of a disbalance in affinity propensities. It is simply the physical result of moving the conductor through the flow of elementary particles.

The movement of elementary particles from the secondary flow into the conductor, however, does deplete the elementary particles in the secondary flow and they have to be replaced from elementary particles that are available, ambient elementary particles that are in use making up the molecules of air, or are satisfying the affinity propensities of lesser emission fields.

Once the elementary particles are captured into the conductor, they have to be accumulated and directed to produce a measurable electrical flow.

This is accomplished by making the conductor a closed path, a circuit around which the elementary particles can travel at their at-rest speed of one hundred and eight-six thousand miles per second, less the physical resistance caused by the conductor.

We can measure the elementary particles that are moving through the conductor by placing a load on the circuit.

A load might be another generator acting as an electric motor, or a light.

The electric motor mirrors the generator, physically reacting to the flows of elementary particles that the generator physically produces.

The light operates on the principle of field replacement, which is the process we have to understand to understand why matter ignites and burns.

The path we have provided for the movement of the elementary particles in the circuit is made out of copper. We have the rheostat on the left, and the copper wire sits on the laboratory bench in front of us.

We could measure the flow in the copper wire by placing a measuring device on the circuit. However, we are going to carry this experiment to the ultimate, and the measuring device will be the copper wire itself.

We turn the lights off and place our hand on the rheostat.

As we begin to turn the rheostat, we increase the flow of elementary particles that is passing through the conductor.

WE CAN CREATE A FLOW OF ELEMENTARY PARTICLES BY ESTABLISHING A PERMANENT POINT OF EXCESS AFFINITY PROPENSITY WITH A DIFFERENTIAL SUFFICIENT TO CAPTURE ELEMENTARY PARTICLES FROM THE LESSER AFFINITY PROPENSITIES IN THE ENVIRONMENT.

As we do so we are also increasing the secondary flow, the inductive flow, the flows that will become so very important when we examine the expansion of light over the expanding sphere of electromagnetic emissions.

For now, we are interested in the primary flow through the conductor.

We slowly move the rheostat so that the flow increases.

The copper wire begins to give off heat, then glow, then turns red hot. The color of the heat proceeds up the spectrum from red to yellow until the copper melts, the circuit is broken, the flows stop, and the color of the melted mass retreats into darkness.

What has happened?

For certified key holders, the copper conductor has melted because too much electricity was allowed to pass through it.

Turn the key in the ignition and the car starts.

This explanation is obviously no explanation at all.

Let's grab some of the conductor's atoms and move them into our work space in non-existence, the area where we perfect our comprehension, so we can isolate just what is going on when something is subjected to the amorphous term "heat".

For our conductor in non-existence, we can use strands of copper atoms, one atom connected to the next.

The ability of the conductor to conduct is found in the nature of the connections between the copper atoms.

In the evolution of the elements, as they broke down from the atom with the greatest number of units in its nucleus, some nuclei would have a greater amount of excess affinity propensity once it had conglomerated all the units it could with respect to its environment than other matter.

The greater the excess affinity propensity that resulted from the particular number of units in the nucleus, the greater the cloud of orbiting elementary particles it would attract.

This condition is repetitive. If you start with a nucleus of say forty units, there might be a large cloud of orbiting elementary particles needed to satisfy the excess affinity propensities of the nucleus. As the number of units was reduced, the excess affinity propensities would diminish, requiring fewer orbiting elementary particles in the cloud. Soon you would reach perhaps twenty-eight units at which point hardly any orbiting elementary particles would be needed to satisfy the excess affinity propensities in the nucleus.

Then, all of a sudden at twenty-four, the greatest number of orbiting elementary particles would again be required.

This repetitive aspect of the number of units as related to excess affinity propensity, and the number of elementary particles attracted into the cloud orbiting the nucleus is what gives the table of elements its periodicity.

THE EXCESS AFFINITY PROPENSITY IS NOT RELATED TO THE NUMBER OF UNITS IN THE NUCLEUS OF THE ATOM, BUT RATHER THE AMOUNT OF AFFINITY PROPENSITY USED UP IN HOLDING THE NUCLEUS TOGETHER. THIS PROVIDES ELEMENTS WITH REPETITIVE CHARACTERISTICS AS THE NUMBER OF UNITS IN THE NUCLEUS INCREASES RESULTING IN THE ABILITY TO ARRANGE THEM ON A TABLE OF PERIODICITY.

The atoms of conductors are among those that need few elementary particles in the cloud of orbiting elementary particles.

When the atoms combine with one another to make up the structures that become matter, they do so by sharing relatively few orbital elementary particles between them.

As a result, the areas between the atoms are relatively free of orbiting elementary particles.

Elementary particles orbiting between the nuclei of atoms making up the conductor physically interfere with the passage of elementary particles tipped into the structure by the process of electric generation, or induced to move through the structure by means of attaching different potentials to the circuit.

The fewer of these elementary particles there are, the freer the elementary particles that have been introduced into the circuit are to travel unhindered at their at-rest motion, the speed of one hundred and eighty-six thousand miles per second.

Modern science likes to think with precision, and therefore puts one electron orbiting the hydrogen atom, two the helium and so forth.

We're going to look closely at our string of copper atoms and realize that even if we could comprehend the number of elementary particles that make up the bindings, the shared orbits holding the atoms together, we couldn't count them because they were moving too fast.

We just see evenly spaced nuclei of copper atoms with a sort of orbital blur between them.

What do we know about the string of copper atoms

that make up our conductor?

We know what we know about any string of any atoms.

The forces that are the properties of the elementary particles that make it up are in perfect balance.

The affinity propensities holding the structure together are exactly balanced by the tendency of the elementary particles that make up the structure to regain their at-rest motion and flash off at the speed of light in combustion.

We know that it is in balance because it will draw enough elementary particles out of the ambient field to balance it, and if it can draw no more, then the presence of the ambient field will balance what is left unbalanced.

THE HIERARCHY OF EXCESS AFFINITY PROPENSITIES IN THE ENVIRONMENT IS AFFECTED BY THE STABILITY OF THOSE EXCESS AFFINITY PROPENSITIES, THEIR PRESENCE.

What is going to happen when we introduce a flow of elementary particles into this perfectly balanced structure?

The flow can physically pass through the structure, although we might expect an orbiting elementary particle to get knocked out of orbit here and there.

It's not the physical result we are looking for.

What do we know about a flow of elementary particles?

It has presence.

What is presence?

At every point in the flow there is an elementary particle because at every point a particle is leaving and another is entering.

This means that at any point in the flow, as the elementary particles move through the conductor, there is an elementary particle.

That elementary particle is stationary with respect to the atoms that make up the conductor.

That elementary particle has an excess of affinity propensity.

Thus, with respect to the nucleus of the atom that is next to the flow, there is an excess affinity propensity that is stationary with respect to that nucleus.

The excess affinity propensity of the nucleus has been used up by attracting elementary particles into paths that orbit the nucleus.

Before the flow was introduced into the atoms, the number of orbiting elementary particles was sufficient to balance the excess affinity propensities of the nucleus.

All of a sudden, there is another source of affinity propensity that can be used to satisfy the excess affinity propensity of the nucleus.

As soon as the flow starts, this source is stationary with respect to the nucleus.

Two elementary particles are not necessary to satisfy a single source of excess affinity propensity.

The nucleus, now having a stationary source of affinity propensity to satisfy its excess affinity propensities, can no longer hold an elementary particle in orbit around it.

The elementary particle departs, moving away from the nucleus it has been orbiting at its at-rest motion of the speed of light.

It might replace another particle physically knocked out of an orbit, it might find an orbit around the flow, it might become a part of the flow, it might leave the structure all together.

THE FLOW THAT PRODUCES PRESENCE IS A DIRECTED FLOW OF ELEMENTARY PARTICLES. ITS AT-REST MOTION IS DIRECTED. AN ORBITING ELEMENTARY PARTICLE IS CAPTURED RATHER THAN DIRECTED. THE PRESENCE OF A DIRECTED FLOW WILL ALWAYS REPLACE THE NEED FOR AN ORBITING ELEMENTARY PARTICLE.

But once a stationary source of affinity propensity becomes available to replace the affinity propensities provided by orbiting elementary particles, the orbiting elementary particles are going to cease to orbit.

They are going to leave their orbits in a flash.

This is field replacement!

Electricity is a directed field. The elementary particles that make up the field, the flows of elementary particles, are directed in a manner that creates a presence, the result of there always being an elementary particle at each point in the flow.

The elementary particles orbiting the nuclei of atoms are captured into an equilibrium by their affinity propensities. Their motion is dictated by the existing affinity propensities. This is also the case of the elementary particles orbiting the directed flows, the inductive flows.

Elementary particles might exist in the ambient field, darting around here and there in response to the changing affinity propensities in the atmosphere.

COMBUSTION IS THE PROCESS OF STATIONARY AFFINITY PROPENSITIES REPLACING LESS STATIONARY AFFINITY PROPENSITIES.

One of the biggest exchanges of ambient elementary particles, for instance, occurs as the day moves into night and then back into day with the temperature changes changing the potentials of the elements in the environment.

As a general rule, directed affinity propensities will replace the affinity propensities satisfied by non-directed elementary particles because of presence.

Presence is stationary. It exists at a particular point in space.

If a directed field is brought into contact with matter, or matter is immersed in a directed field, the affinity propensities of the directed field will replace the affinity propensities binding the matter together, allowing the elementary particles that were providing those affinity propensities to depart at their at-rest motion, the speed of light.

Now we can see, if we go back to the conductor, what is going to happen as we move the rheostat so that the flow of electricity increases.

As the flow increases, presence increases, and with the increase of presence, the stationary affinity propensities allow more and more of the orbiting elementary particles to cease their orbiting.

During the first portion of the increase, the released elementary particles go into making up the inductive flows needed by the primary flow of electricity. Up to a point this is a complimentary process, with the elementary particles in the inductive flows moving back into the structure and becoming orbiting elementary particles as soon as the flow ceases and the stationary affinity propensities associated with the flow disappear, leaving the nuclei with an excess of affinity propensity.

But, if we continue to turn up the juice, sooner or later the field will be so strong as to cause orbiting elementary particles to be released, elementary particles that have no use in the structure as it exists with the flow.

These elementary particles will leave the structure, evidencing themselves as heat.

If we relentlessly turn up the flow, the point will be reached where the number of elementary particles that are trying to leave the area exceeds the number that can leave the area.

These elementary particles then start to organize their departure in the orderly manner that results in an emitted electromagnetic spectrum, light.

If we pump up the volume even more, then the elementary particles that bind the atoms together in a structure will cease to serve that purpose, and the atoms will start to change position with respect to one another, resulting in a noticeable change in the configuration of the physical matter.

THE STATE OF MATTER CAN BE DEFINED BY THE AFFINITY PROPENSITIES WITH WHICH IT IS INVOLVED.

It melts!

Field replacement, the exchange of a more stationary affinity propensity for a less stationary affinity propensity so that the elementary particle providing the less stationary affinity propensity will regain its full at-rest motion and depart the structure is more dramatically demonstrated in the process of using one match to light another.

Because we are going to look closely at how the second match ignites and begins to undergo combustion, we will start with a lighted match.

The match is emitting streams of elementary particles.

The elementary particles are leaving the physical structure at such a rate that they have to line up and await their turn.

This alignment creates the structure that we measure as the electromagnetic spectrum, light.

While individual streams are being emitted, and breaking down into smaller streams as they move away from the match, the streams can be visualized as a sphere which at each instant forms at the edge of the combustion process.

After one second of formation, the sphere, if unhindered, would expand to have a radius of one hundred and eighty-six thousand miles.

> **LIGHT CAN BE CONSIDERED AS A SERIES OF EXPANDING SPHERES WHOSE SURFACES ARE CONNECTED BY ELEMENTARY PARTICLES, WHICH IN TURN CAN BE CONSIDERED LINEARLY AS STREAMS OR FLOWS.**

However, because there is a new sphere emitted each instance the match is undergoing combustion, we can consider that there is a sphere of light around the burning match. For our purposes, the effective sphere might be three feet, the distance the match lights up the surrounding area. After that distance, the light begins to break down and the elementary particles that made it up are absorbed into the changing affinity propensities of the atoms and molecules that make up the environment, the air.

At any point in the sphere of light, there is presence,

the existence of the flows of elementary particles.

Where there is presence, there are stationary affinity propensities.

The stationary affinity propensities in the sphere of light being emitted by the burning match are proportional in strength, increasing or decreasing inversely with the square of their distance from the match, depending on which way you are moving in the emitted field.

If you put your finger a foot away from the match, you can't feel anything.

If you put it an inch away from the match, you might feel quite a bit of the emitted particles as they bounce off or flow through the surface of your skin in the form of heat. If you put your skin close enough, the field replacement will occur in the skin. You reach a point in which the stationary affinity propensities of the emitted field become stronger than the orbiting elementary particles that are holding the molecules of the skin together, and they depart, leaving you with the matter that makes up your skin changed, perhaps as a blister.

When we have a sphere of electromagnetic radiation, light, what we really have is a sphere of stationary affinity propensities, with the strength of the excess affinity propensities available to replace less stationary affinity propensities becoming increasingly strong the closer they are to the source of the light, the increase occurring with the square of the distance to the source.

With one match lit, we can take the head of an unlit match and begin to move it into the emitted field of the lit match.

The atoms of the elements that make up the head of the match are held together in equilibrium by sharing orbiting elementary particles.

The orbiting elementary particles do not produce stationary affinity propensities.

The field of the lit match does provide stationary affinity propensities.

AS MATTER MOVES INTO AN EMISSION FIELD, THE FIELD GROWS STRONGER INVERSELY WITH THE SQUARE OF ITS DISTANCE FROM THE SOURCE.

We are providing the physical motive power that is moving the match deeper into the field. As we do so, we are causing the less stationary elementary particles in the unlit match to begin to move out of their orbits. Compared to the copper conductor, however, the number of elementary particles orbiting in clouds is vast.

And unlike the flow through the conductor, the flows that make up the light have already captured all the elementary particles they need to make up their inductive, or secondary flows.

The match head will start to heat up as the elementary particles replaced by the stationary affinity propensities in the lit match's field begin to be released into the surrounding field.

But this isn't ignition. The match head hasn't yet begun to undergo combustion without reference to the lit match's replacement field.

What is ignition?

As the orbiting elementary particles are released in the process of field replacement, they reach the point at which there are more of them being released than can leave the surface of the match.

IGNITION IS THE POINT AT WHICH THE ELEMENTARY PARTICLES HOLDING THE ATOMS TOGETHER ON THE SURFACE OF MATTER BEGIN TO BE EMITTED AT A RATE THAT CREATES A DIRECTED FIELD WITH AFFINITY PROPENSITIES GREATER THAN THE AFFINITY PROPENSITIES ON THE SURFACE OF THE MATTER.

When this occurs, the affinity propensities within the physical structure of the match increase independent of the replacing field. As the elementary particles are released and can't leave the area, their affinity propensities now go into replacing the affinity propensities of additional orbiting elementary particles.

The point is quickly reached where the elementary particles have lined themselves up for an orderly release.

When they do so, they burst forth from the surface of

the match in the form of the electromagnetic spectrum.

The match ignites!

Once combustion occurs, the process of field replacement begins to occur within the match head, and the match burns as long as it produces the stationary field that will replace the orbiting elementary particles in the matter that is making it up, allowing the release of those particles into the field at their at-rest motion, the speed of light.

A light bulb, operating in a vacuum, performs the unique task of converting the elementary particles that make up the flow of electricity directly into an emission field.

> **WHEN WE MANIPULATE ENERGY, WE ARE ACTUALLY MANIPULATING THE AFFINITY PROPENSITIES IN MATTER SO AS TO CREATE FLOWS OF ELEMENTARY PARTICLES TRAVELING AT THEIR AT-REST MOTION, THE SPEED OF LIGHT.**

This is accomplished by pouring streams of elementary particles into a load, a point of resistance, so that there are more elementary particles in the matter providing the resistance than can be emitted from the surface of the matter. This causes the ordering effect of the elementary particles at the surface of the matter creating the resistance, and the resulting emitted electromagnetic spectrum.

We can picture a room with a single light, the light illuminating a generator, the generator producing the electricity that flows through the light.

There is one circuit in the room, the obvious one that connects the generator to the light, the generator creating an electric current that flows through and lights up the light bulb.

This is the key circuit, the "turn the key and start the car" circuit.

It takes a certain level of knowledge to understand how the generator works, how to wind the coils, how to place the brushes, how to wire the system. It also takes a great deal of knowledge to fabricate the light bulb, more than the basic knowledge of how to screw it into the socket, and it takes a level of knowledge to know how to connect everything up and put in a switch so the whole thing is workable.

There is a level of knowledge associated with the obvious circuit.

However, there is another circuit in the room.

The windings on the rotors are arranged so that they not only produce a secondary flow of elementary particles as they rotate, they knock particles out of those secondary flows so that a primary flow of elementary particles is established in the circuit.

The secondary flows, having lost particles to make up the primary flow, have to use the affinity propensities being created to obtain the elementary particles out of the ambient field, the field of randomly moving elementary particles that make up the environment, or perhaps the orbiting elementary particles that hold the atoms of air together, or the water moisture in the air.

A HYPOTHESIS THAT THE EARTH HAS AN EMISSION FIELD MIGHT EVOKE AN EXPERIMENT IN WHICH A RADIO ANTENNA COULD BE USED TO DETECT THE STEADY NOISE THAT SUCH AN EMISSION FIELD WOULD PRODUCE. IF THAT NOISE WERE UNIFORM OVER THE SURFACE OF THE EARTH REGARDLESS IN WHICH DIRECTION THE ANTENNA WAS POINTED, THEN IT COULD BE CONCLUDED THAT THE ANTENNA WAS DETECTING THE EMISSION FIELD. THE EXPERIMENT MIGHT BE CARRIED FURTHER BY SHOOTING A DETECTOR SIX HUNDRED MILES FROM THE SURFACE OF THE EARTH TO SEE IF DISTANCE MADE THE DETECTION OF THE EMISSION FIELD LESS UNIFORM.

As these elementary particles are cycled from the ambient field into the secondary flows, and then into the primary flow, they establish a current in the conductor. The current, which is a flow of elementary particles, moves through the conductor to the light. At the light, it hits the resistance of the load, the light's filament, and the elementary particles that make it up begin to collect at that source, backing up until they can order themselves in a manner that will allow them to be emitted from the filament.

As soon as they are emitted, they become a part of the electromagnetic spectrum that makes up the expanding sphere of light surrounding the light bulb.

The flows that make up this expanding sphere bounce off the walls, interact with the molecules of air, and generally begin to break down.

As the flows break down, the elementary particles that made them up begin to become part of the ambient field of elementary particles that fills the room.

As part of this ambient field, they once again are captured into the secondary flows, become a part of the primary flow, and race off at their at-rest speed to the filament to once again be emitted as light.

SOMETIMES WHAT WE SEEK IN THE DEPTHS OF INTERSTELLAR SPACE IS RIGHT BENEATH OUR FEET.

This second circuit is another level of understanding, a level beyond simply turning the key, throwing the switch!

However, the generator is usually collecting elementary particles and sending them far away to operate loads in peoples houses and businesses.

The ambient field from which the generator gets its elementary particles is not limited to the elementary particles that are in the air and moisture, and broken down light in the room in which the generator operates.

The generator has a far more potent source of elementary particles.

That source is located some hundred miles beneath the

generator's floor, on the surface of a contracting sphere of combustion, a sphere that is itself producing a continuous expanding sphere of electromagnetic radiation, the Earth's combusting interior.

While meteorological and geophysical ruminations require an Earth that is in heat balance, and gives up the precise amount of heat it receives from the sun, geology recognizes an outward flow of heat from beneath the Earth's crust into space.

Geologists estimate that this dissipated energy, the emission field of the cooling Earth, is in the area of six trillion calories per second, a meaningless figure, but a significant fact.

8. THE EXPANDING SPHERE: THE ATTRACTIVE FORCE

It is hard to assess the damage to the effort to develop a consistent picture of physical phenomena that the analogy of light to water waves has caused. It is on a par with the blind devotion to the mass/gravity concept in its ability to obscure reality.

The analogy has its basis in the fact that light creates interference patterns. Interference patterns, a misnomer in itself, is the physical result of dividing light from the same source and intermingling the resulting flows.

THE ANALOGY OF LIGHT TO WATER WAVES RIVALS THE MASS/GRAVITY CONCEPT FOR ITS ABILITY TO OBSCURE REALITY.

If you take a source of light, pass it through two vertical slits in a screen, the light that falls on a solid screen placed in front of the screen with the two slits is barred. There are vertical bars of light alternating with vertical bars of no light.

If the distance of the two slits is adjustable, and you move one of the slits forward or backward, the barred effect will cease to exist.

Now, if you take a pool of water and drop two stones in at different points, the waves that move out from each point will eventually intersect. When they do so, there will be points at which the trough of one wave will equal the peak of another wave, and the result will be no wave.

No wave in water?

No light on the collecting screen?

Gosh. The two effects are identical.

Lo, light must travel in waves.

On such conceptual foundations rests the edifice of modern science.

SINCE LIGHT HAS BEEN FOUND TO BEHAVE LIKE A PARTICLE, ELECTRONICALLY IMAGING PICTURES IN VIDEO CAMERAS, AND SINCE IT IS KNOWN TO BE A WAVE, MODERN SCIENCE HAS CONCLUDED THAT LIGHT IS A WAVE PARTICLE, OR IN CERTAIN CASES, A PARTICLE WAVE.

Well, ancient science also, because water waves had to have a medium (not to mention gravity!). Much effort prior to the current century was spent in attempting to use the medium that light had to have to travel as waves to discover the absolute motion of the Earth, another endeavor, like figuring out how much the Earth weighs, that seems to occupy so much of our learned time.

During one of the experiments that were designed to determine the absolute motion of the Earth, it was discovered that time and distance weren't measurements, but rather quantities that themselves could be measured. This led to the current universe where mass increases with speed, time shortens, and the universe closes in on itself so that if you go in one direction far enough and fast enough you'll end up where you started without having aged one second give or take a tick.

All of these things have been proved beyond a reasonable doubt, leaving nothing about physical reality to analyze, except perhaps what makes the sun come up every morning.

Most technology dealing with light has to use as its assumption that light is made up of discrete particles, and this is indeed the case with our elementary particle, light being composed of flows of elementary particles traveling at their at-rest speed of light after being released by the combustion process from the matter they make up.

QUANTIFYING THE TOOLS CREATED TO ANALYZE PHYSICAL REALITY RESULTS IN UNREALITY.

Most technology today has to deal with the conversion of light to electricity and vice versa, which is what we also are doing with the elementary particle we have synthesized.

Thus, I think we can forget the analogy to water waves.

Or at least we could if we could show how light, as discrete particles, produces interference patterns.

The interference patterns are not only explainable in terms of light being structurally composed of elementary particles, they are the clue that explains how light expands uniformly over the surface of the expanding sphere.

If you want to understand a physical effect, you have to understand the physical processes that are going on that result in the physical effect.

The light that is going through a slit is a portion of light that is expanding over the surface of an expanding sphere.

What happens as soon as it passes through the slit?

It begins to expand once against over the surface of an expanding sphere.

INTERFERENCE PATTERNS ARE NOT THE LINEAR RESULT OF THE DISTANCE OVER WHICH LIGHT HAS TRAVELED, BUT RATHER THE WAY LIGHT HAS BEEN EXPANDED OVER THE SURFACE OF THE EXPANDING SPHERE THAT MAKES IT UP.

The slit provides a second focal point from which the light expands.

This should be fairly obvious.

We have one expanding sphere of light.

We make the light go through a hole.

The light on the other side of the whole is still expanding light.

The hole, however, is now the center of its expansion.

We thus have created a second expanding sphere of light with a new distance to its source.

The light that makes up the new expanding sphere of light will begin to diminish at a specific rate, inversely with the square of this new distance.

At any point in the new expanding sphere of light, we will be able to determine the presence of that light by reference to this new distance.

What we have done in essence is to overexpand part of the light that made up our original expanding sphere of light by providing a new area into which it could expand.

But we haven't created just one new expanding sphere of light. With two slits, we have created two expanding spheres of light.

THE LIGHT COMING THROUGH THE TWO SLITS CONSTITUTES TWO EXPANDING SPHERES OF THE SAME LIGHT WITH A DIFFERENT FOCAL POINT. THE AMOUNT OF LIGHT EXPANDING IN THE NEW SPHERES IS PROPORTIONALLY LESS THAN HAD THE LIGHT NOT BEEN EXPANDED FROM THE NEW FOCAL POINT.

As long as the slits are the same distance from the original source of the light, then the amount of light that is being overexpanded by passing through the two slits is identical in each of the two new expanding spheres of light.

As that is the case, then at the identical point in each of the spheres as determined by the new distance to the focal point of expansion, there is an identical amount of presence.

What do we know about any point in the expanding sphere of light?

If there is presence, then there is always a particle

entering and leaving at any point, and at any point in the stream there is an affinity propensity.

The streams of elementary particles, having presence, cause elementary particles to orbit the flow at right angles.

The light thus has an inductance associated with it as it expands.

What happens when the two expanding spheres, containing corresponding points with identical presence, are intermingled?

WHEN THE FLOWS OF THE NEW EXPANDING SPHERES ARE INTERMINGLED, THEY RECOMBINE SO THAT THERE ARE AREAS OF LIGHT AND NO LIGHT. THE QUESTION IS, WHAT CAUSES THE FLOWS TO RECOMBINE.

In the laboratory, if you place one conductor next to another and pass an electric current through both in the same direction, the inductances will cause the wires to move together.

The inductances will combine the flows.

This is what is happening when the two over-expanded spheres of light are intermingled one with the other.

The inductances re-combine the flows. When the light falls on the screen, there is simply no light between the places where the recombined light strikes the screen.

The notion of light waves was that they canceled each other out when they were split and arrived at the screen at the same time. If you altered the distance of one of the waves to be intermingled, then they didn't arrive at the screen at the same time and the interference patterns didn't appear.

If an experiment could be devised which would alter the distance by manipulating the other terms of the rate times time equals distance equation we have already seen is the equation that is used to measure physical reality, then we should be able to eliminate the interference patterns.

The Michelson-Morley experiment, while designed to determine the absolute motion of the Earth, inadvertently did exactly this, and when it was found that the interference patterns still existed, assumptions were made about the rate altering the time and distance that the light traveled.

WE CREATE DETECTORS TO REVEAL SOMETHING ABOUT PHYSICAL REALITY, MAKE UNCONTROLLED ASSUMPTIONS ABOUT WHAT IS BEING DETECTED, AND THEN PROVE THE ASSUMPTIONS BY REFERENCE TO THE DETECTOR.

The faulty concept of light as a wave resulted in the use of a detector in creating assumptions about space and time that have since continually been proven by reference to the detector, the Michelson-Morley experiment.

However, as the interference patterns result from the geometric nature of light's expansion, rather than the rate times time equals distance formulae, the results of the

Michelson-Morley experiment are to be expected. Light is divided, expanded and re-expanded so that when the separated flows are intermingled, the presences are identical with respect to both expanding spheres. This allows the inductances to recombine the flows so that the interference patterns appear as expected.

The interference patterns do not result from distance, but rather from the physical relationship of the expansion of light with respect to the distances over which it has expanded, a relationship that is not affected by the motion of the source of the light.

IF INDUCTANCES RECOMBINE OVEREXPANDED LIGHT TO PRODUCE INTERFERENCE PATTERNS, WHAT EFFECT DO THEY HAVE ON LIGHT THAT IS SIMPLY EXPANDING?

Of course, the existence of interference patterns taken in conjunction with the way that they are produced, accords with the concept of an elementary particle with two properties, an at-rest motion that is the speed of light with respect to a stationary conglomeration of elementary particles, and affinity propensity, the propensity to conglomerate together as each tries to occupy the space occupied by the other.

Interference patterns are what we would expect from flows of elementary particles if those elementary particles were the same particles that produced electricity and the inductances associated with it.

But the ability of inductances to recombine flows of elementary particles has far reaching consequences when it comes to the effect of that recombination on the units that make up the nucleus of an atom.

We tend to hear without ears the statement that light expands uniformly over the surface of the expanding sphere.

This statement is derived from the belief that light diminishes inversely with the square of its distance from the source.

Because modern science has no concept of what energy is, it sort of just gets uniformly less as it gets further from its source.

We have no way of directly measuring the proposition, but we don't have to so long as it is expanding over the surface of the expanding sphere that is following geometric rules that are measurable.

But does light expand uniformly?

We have already noted in passing that combusting objects do not cool in a straight line.

They oscillate between heating and cooling.

Just as something doesn't cool in a straight line, so the emissions it creates don't expand uniformly.

Light expands into the space available for its expansion, the spherical area around its propagation, the release from the matter undergoing combustion of the elementary particles that make up the matter.

When light's focal point changes to create interference patterns, it overexpands.

When light in an expanding sphere of electromagnetic emissions reaches any point of expansion, when the flows break down into smaller flows, it also overexpands, the smaller flows overexpanding from each new point of expansion.

Another way of characterizing overexpansion is to say that the light's presence diminishes to an extent greater than the inductive forces its presence creates at any point in the flow.

LIGHT DOES NOT EXPAND UNIFORMLY. IT EXPANDS AND CONTRACTS AS THE FLOWS BREAK DOWN TO UNIFORMLY COVER THE SURFACE OF THE EXPANDING SPHERE.

When overexpansion occurs, the inductive forces that exists at any point in the expanding sphere are sufficient to recombine the flows.

The recombined flows, moving forward at the rate of one hundred and eighty-six thousand miles per second, immediately overexpand and are immediately recombined by the inductive flows.

This is occurring simultaneously at every point in the expanding sphere of electromagnetic emissions.

In short, the expansion process is a process of expansion and contraction.

The particles that make up the streams are moving forward, expanding outward.

The presence and the inductance the presence creates are stationary at any point in the flow, and is determined by the geometric nature of the expanding sphere of light.

The forward moving streams are continually expanding into inductances that recombine them, the inductances determined by the distance from the source of the emissions that the flows are continually outracing.

At any point in the expanding sphere of electromagnetic emissions, the expanding light is undergoing a retrenchment of the expansion process, a process in which the flows are momentarily recombined.

THE EXPANSION LEG OF THE EXPANDING SPHERE OF LIGHT WOULD HAVE NO EFFECT ON THE UNITS THAT MAKE UP THE NUCLEUS OF AN ATOM. THE CONTRACTION LEG WOULD HAVE THE EFFECT OF PHYSICALLY MOVING THE UNIT BACK TOWARD THE SOURCE OF EMISSIONS.

This retrenchment is actually a mechanism.

The mechanism can be visualized by holding one hand open and then bringing the tips of the fingers and thumb together in one motion.

The fingers in the open position represent flows of expanding light.

As they come together at the tips, they represent the

effect of the inductive flows on the expanding flows.

The light flows expand over the area available. The expansion causes them to break apart into smaller flows. The smaller flows intermingle with the surrounding flows. The inductances cause the flows to recombine. As they recombine they begin to expand, break apart, intermingle, recombine, expand, break apart, intermingle, recombine, expand and so forth.

It is the recombination of the overexpanded flows that affects the units at the nucleus of the atoms that exist in the field of expanding emissions.

The recombination is a physical process and it occurs in place on a repeated basis.

It grabs the unit of a nucleus of an atom and propels it back toward the source of the flows.

The direction is established by the point where the recombination begins.

The point furthest into the expansion is the point at which the flows to be recombined have the least presence and where the number to be recombined is the greatest.

The mechanism of recombination begins at this point, rather than at the point closest to the source of expansion where the flows are still just beginning to break down and therefore are still comparatively strong.

The mechanism acts on the individual unit in the nucleus of an atom, so all units are propelled toward the source of the emissions at the same rate.

However, if you attempt to move an atom against the recombining mechanism, every unit that is tied together by its internal affinity propensities will have to be moved at the same time so that an atom with five units will have to be moved against the force of five recombining mechanisms, one with a hundred units tied together will have to be moved against a hundred recombining mechanisms.

MATTER, MOVING IN RESPONSE TO THE MECHANISM THAT REGULATES THE EXPANSION OF LIGHT, WOULD MOVE IN ACCORDANCE WITH THE INVERSE SQUARE LAW MEASURED FOR GRAVITY BECAUSE THE STRENGTH OF THE MECHANISM, BASED ON THE PRESENCE OF THE EXPANDING ELECTROMAGNETIC SPECTRUM, IS INVERSELY PROPORTIONAL TO ITS DISTANCE FROM THE SOURCE OF EMISSIONS.

Thus, all matter moves toward the source of emissions at the same rate, but takes differing forces, depending on the number of units tied together in the nucleus of its atoms, to move against the source of emissions.

And what is this rate?

At any point in the expanding sphere, the recombining mechanism exists, it is stationary, and it exists in proportion to the presence at that point.

The flow's presence creates the orbiting elementary particles.

It is these orbiting elementary particles that at each

point in the expansion process recombine the intermingling flows as they overexpand.

The force of the recombination is therefore directly proportional to the secondary flows of elementary particles, the inductances that recombine the flows.

The secondary flows are proportional to the flows.

The flows are proportional to their distance from the source.

The proportion is inverse with the square of the distance.

Thus, the force of the mechanism recombining the expanding flows of light diminishes inversely with the square of its distance from the source of the emissions.

If a unit of an atom was caught in one recombination mechanism, it would be pushed back toward the source of the emissions, which in turn would subject it to the next closest recombining mechanism which in turn would push it back toward the source which in turn would subject it to the next closest recombining mechanism and so forth.

The unit would thus move toward the source of emissions at an accelerating rate, the rate of acceleration being that it would increase with the square of the distance over which it moved.

And if you don't think that the units that make up the nucleus of an atom don't move toward the source of electromagnetic emissions, just try to keep the dust off your television screen.

9. THE EXISTENCE OF MATTER: THE FIELD FACTUAL CONGRUITY

Light is the electromagnetic spectrum. As light expands, the inductive flows established by its presence create an interplay between expansion and contraction. This interplay results in the attractive mechanism. The attractive mechanism results from the process of recombining the intermingled streams of light that have overexpanded.

The inductive flows recombine the expanding flows. In the process, the mechanism physically acts on the units of elementary particles that make up the atoms of matter. It causes them to move back toward the source of the flows of light, the emitter that is undergoing combustion.

THE MOVEMENT OF MATTER IN SPACE IS THE RESULT OF THE PHYSICAL MECHANISM THAT REGULATES THE EXPANSION OF LIGHT PHYSICALLY ACTING ON THE PHYSICAL UNITS THAT MAKE UP THE NUCLEI OF ITS ATOMS.

The attractive mechanism, acting on the individual units, moves the units toward the emitter at an accelerating rate. This rate accelerates them inversely with the square of the distance traveled. This is because the strength of the mechanism diminishes inversely with the square of its distance from the emitter.

However, the units of the atoms are bound together by their internal affinity propensities. As a result, movement away from the emitter involves moving all the units against each of the mechanisms acting on each of them. It thus takes an effort proportional to the number of units in the nucleus of the atom to move it against the attractive mechanism.

This is what we measure as weight.

The atom, as it accelerates in the expanding field of light, is moving with the total force of the re-combinations on each of the units of its nucleus. This means that atoms with a different number of units in the nucleus will fall at the same rate. When the same atoms begin to move against the expanding field, the bindings holding the units together in the nucleus come into play. At that point the force of the movement is the total force that was applied to move the individual atoms when they were moving toward the emitter.

INERTIA AND MOMENTUM ARE NOT PROPERTIES OF MATTER BUT THE PHYSICAL RESULT OF THE INDIVIDUAL UNITS OF THE ATOMS AT REST WITH, OR MOVING WITH OR AGAINST THE PHYSICAL ACTION OF THE ATTRACTIVE MECHANISM.

This is what we measure as momentum.

There is no mass.

There is no density, other than the fact that some atoms have more units in the nucleus than others, and some atomic and molecular structures occupy less space per volume than others.

Atoms, and structures of atoms respond both to the affinity propensities in the environment, and to the attractive mechanism as it exists in expanding fields of electromagnetic radiation.

As the attractive mechanism moves an atom deeper into the field, the field replacement increases proportionately.

Matter moves to find its position with respect to the current forces acting upon it. These forces are the attractive mechanism that causes it to move, and the affinity propensities that cause it to undergo combustion. The former causes matter to move deeper into combustion fields which in turn increases the elementary particles available for combustion.

If we light a match, we can burn paper with it, the paper that makes up the body of the match, or a piece of newspaper that we want to use to build a fire.

If we have laid the fire right, then we can use the paper to light the kindling, which in turn will light the larger sticks that can in turn light the logs that make up the fire.

In each stage, we are creating a stronger field to start the combustion process in matter that is more stable with respect to the equilibrium of the affinity propensities that make it up.

If we cook atoms and molecules over a bunsen burner, we are replacing the existing balance of affinity propensities with the affinity propensities of a directed flow of elementary particles.

When we temporarily replace orbiting elementary particles that are providing the equilibrium that is holding the atoms and elements together in some specific physical construction with a directed field, we strip the structure of these orbiting elementary particles. This allows them to become part of the directed field, or escape altogether in the form of heat.

WE CREATE CHEMICAL REACTIONS IN MATTER BY FIELD REPLACEMENT, USING A DIRECTED FIELD TO ALLOW ATOMS TO RELEASE ORBITING ELEMENTARY PARTICLES, PHYSICALLY REPOSITIONING THE ATOMS, AND THEN REMOVING THE DIRECTED FIELD SO THAT THE NEW ARRANGEMENT OF ATOMS CAN PULL ELEMENTARY PARTICLES OUT OF THE AMBIENT FIELD TO BALANCE THE NEW ARRANGEMENT OF AFFINITY PROPENSITIES.

If we add other atoms or molecules after we have stripped the existing atoms and molecules, and remove the directed field, the disbalances created by withdrawing the directed field will create excess affinity propensities. These excess affinity propensities will result in the atoms and molecules that are now physically contiguous with one another recombining to regain equilibrium. In the process of recombining, another specific physical structure will develop.

We can crack oil, and then use it to make all sorts of

other useful products, as well as power our own motion.

We can make an arc that will cut the atoms of steel, and we can create an atomic explosion that will separate them altogether.

If an asteroid should get caught in the expanding field of light emitted from the sun, and be pulled toward it by the increasing strength of the attractive mechanism, it wouldn't make much difference what the asteroid was made out of. It would start to get warm as it progressed into the field. It would start to become hot and emit light as it ignited and began to undergo combustion. By the time it reached the surface of the sun, the atoms that made it up would have been broken down into their individual units. The individual units, when they reached the sun's surface, would begin to come apart as the individual elementary particles that made them up regained their at-rest speed.

What does this tell us about matter and the field that it occupies?

Matter exists with respect to the field that it occupies!

The stronger the field of light, of electromagnetic emissions, the more presence that field will have. The more presence, the more the field will be able to replace the elementary particles that make up the matter. The replaced elementary particles will in turn regain their at-rest motion of the speed of light and depart the structure.

This means that, given a particular field, some matter

couldn't exist in that field because it would have broken down into something else. Other matter could exist in the field, but only marginally because it would be in the continual process of breaking down. And some matter could exist in a very stable form.

We have to recognize that there are margins of stability in the existence of matter, just as there are margins of stability involved with objects at rest in a field of attraction.

BECAUSE AFFINITY PROPENSITIES IN THE EXPANDING SPHERE OF ELECTROMAGNETIC EMISSIONS REPLACE THE AFFINITY PROPENSITIES HOLDING MATTER TOGETHER, THE EXISTENCE OF MATTER IS ALWAYS RELATIVE TO THE STRENGTH OF THE EMISSION FIELD THAT IT OCCUPIES.

Just as it may take a super shove to get the car rolling down the driveway, once all the units that make up all the atoms of the car have moved against the attractive mechanisms that are holding them in place against the matter that separates them from the source of the expanding field of electromagnetic radiation giving rise to the attractive mechanism, once this inertia is overcome, then the car moves easily to the gas station.

With respect to the balanced affinity propensities that make up atoms, and groupings of atoms, it will always take a little more affinity propensity to break the existing equilibrium than exists in holding it together.

Thus, there are tolerances within which matter can exist in a field. The best example is water. It exists between the field at which it boils, and the field at which it crystallizes and becomes ice. The range of this field is quite large, by design, a hundred degrees on the centigrade scale.

If there was water on a planet as far away from the sun as Pluto, it would be in the form of ice. Water in liquid form, or in its gaseous state of steam or water vapor, would be non-existent. The sun's emission field would be too distant to convert it from the crystalline form.

JUST AS H_20 MIGHT BE ICE, WATER OR NON-EXISTENT DEPENDING ON ITS DISTANCE FROM THE SUN, SO ATOMS WILL BE STABLE, RADIOACTIVE OR NON-EXISTENT DEPENDING ON THE EMISSION FIELD THAT THEY OCCUPY.

The Earth is deeper in the sun's emission field, and the planet is considered to be the water planet, covered with H_2O although it also has abundant ice and water vapor.

But water on Mercury would soon be vaporized, at least at its equator. Mercury is too deep in the sun's emission field to allow water to exist either in its crystalline or liquid form.

If we examine the formation of an atom in non-existence, we can define non-existence as the lack of a field.

In non-existence, there is no emission field.

We created non-existence to show how an atom might form out of elementary particles with at-rest motion and affinity propensity.

We know that the combustion process breaks down the atoms into their most basic elementary particles. We have seen that those elementary particles are emitted in the electromagnetic spectrum.

Thus, matter can not continue to exist in the strongest field, the combustion field that exists at the surface of a combusting star such as the sun.

There is no reason why the opposite isn't also a consistent description of physical phenomena.

Matter forms in the absence of a field.

This would provide a universe in which matter was continually forming in areas where there was no field, in which the emission fields from expanding spheres of light were breaking down into individual elementary particles. Matter would continue to form until it ignited by the process of field replacement, at which point the matter would begin to undergo combustion. Combustion would create once again expanding spheres of light that would expand, break down, and cycle through the process still one more time, and a time after that, and after that.

This is actually a five part process, where the elementary particles form into matter, the point at which the affinity propensity gains ascendancy over the at-rest motion and is dominate; The matter ignites, the point at which the at-rest motion begins to regain ascendancy, the point at which the at-rest motion is ascendant but the surface area is not great enough for all of the elementary particles to depart the structure; it undergoes combustion, which results in the expanding sphere of electromagnetic radiation which is the orderly result of the inability of the elementary particles to exit, a structure that balances the affinity propensities and at-rest motion in a structure that expands uniformly over the area available for its expansion; And the light expands to the point at which it breaks down and the at-rest motion of the particles takes over, resulting in random motion and the resulting collisions of elementary particles that marks the beginning of matter formation.

THE ATTRACTIVE MECHANISM, EMBEDDED IN THE EXPANDING SPHERE OF ELECTROMAGNETIC EMISSIONS ENSURES THAT ALL MATTER WILL PARTICIPATE IN A CONTINUAL CYCLE OF FORMATION, COMBUSTION, DISSIPATION AND REFORMATION.

The expanding sphere of electromagnetic emissions contains the attractive mechanism. This mechanism is stationary and operates on the units of the nuclei of the atoms to move them toward the combusting source of the emissions to ensure that the process involves recycling all matter in the five-part process.

If we put the matter that forms in the lack of a directed field at one end of the spectrum, the most complex atom field at one end of the spectrum, the most complex atom possible with the most number of units in the nucleus, and the matter that can only exist in the strongest field possible at the other end of the spectrum, the field that exists at the surface of the sun, a single unit of matter that measures out as the nucleus of a hydrogen atom, then we can get some sort of idea what the field factual congruity is.

The Earth's emission field is a combination of the expanding field of electromagnetic emissions that originate some hundred miles below our feet and the more obvious emission field of the sun.

ATOMS SUCH AS URANIUM WHICH ARE RADIOACTIVE IN THE EARTH'S EMISSION FIELD EMIT ELEMENTARY PARTICLES IN AN EFFORT TO REACH STABILITY IN THAT FIELD.

If we use the uranium series as an example of matter existing in our field with the greatest number of units in the nucleus, we find uranium existing in equilibrium with other elements or compounds of elements.

However, if we separate the uranium out from the elements that are providing the uranium atoms with the affinity propensities that allow them to remain in equilibrium in the Earth's emission field, then the Uranium atoms become emitters.

Alone, these atoms are too complex to exist in the Earth's emission field. The elementary particles that bind the atoms together, and the elementary particles that are involved in holding the units together in the nuclei begin to depart the atoms as a result of field replacement.

This is because the strength of the field of electromagnetic emissions on the surface of the Earth exceeds the ability of the uranium atoms to exist. As the field fluctuates, it replaces the affinity propensities in the atomic structure of the uranium atom and elementary particles are emitted.

The uranium breaks down and seeks other elements and compounds with which it can combine to once again become stable. This results in the phenomena of a series of related radioactive arrangements, as the matter attempts to regain equilibrium.

If we discount the emission fields of the bodies on which the elements are found, we could state that uranium, ninety-three million miles from the surface of the sun, exists as a boundary element.

That means that the nucleus of the atom contains too many units for the atoms to continue to exist in the field. It will slowly breakdown into atoms with fewer units in the nucleus. This will continue until it regains its internal equilibrium, the balance of the affinity propensities of the elementary particles making up the structure of the atom with the affinity propensities of the directed field in which the atom finds itself.

If uranium is a boundary atom ninety-three million miles from the surface of the sun, what type of atom might it be if it found itself as close to the sun as the orbit of

Mercury?

It couldn't exist in the orbit of Mercury. The field would be so great that it would have broken down into atoms with fewer units in the nuclei.

What would happen if we took uranium atoms and moved them to the orbit of Mercury?.

They would undergo the combustion process, heating up, beginning to emit in the visible spectrum, if there were enough of them, looking like a miniature star, the atoms breaking down into the individual units on the surface of the combustion process. We would measure them just as we measure the surface of the sun, as hydrogen atoms.

URANIUM MIGHT NOT BE RADIOACTIVE ON PLUTO, THE AFFINITY PROPENSITIES IN THE FIELD FAR FROM THE SURFACE OF THE SUN BEING INSUFFICIENT TO AFFECT THE ATOM'S STABILITY.

What if we took the uranium atoms and moved them out to the orbit of Jupiter?

The field would be so weak as to allow the atoms to exist.

Not only could they exist, however, they would not be radioactive. The sun's emission field in the orbit of Jupiter is too weak to replace the internal equilibrium of the elementary particles that make up the uranium atom.

Would there be a radioactive atom in the orbit of Jupiter?

We have created on Earth a number of elements that can only exist at the moment of their creation because the field on the surface of the Earth is so strong that the atoms immediately break down.

These elements all have nuclei with more units in the nucleus than uranium has.

Perhaps these elements would exist on Jupiter if they could find some elements and compounds of elements to place them in radioactive equilibrium. If there were Jovians, perhaps they too could extract these elements from their equilibrium, turn them into emitters, and devise a way to create combustion fields that would also be too great for them to exist in, just as we have.

What would the hypothetical Mercurians use as a boundary element to make an atom bomb? Perhaps iridium, stable on Earth, but possibly a boundary element on Mercury.

Which is to say that because elements exist with respect to the field in which they find themselves, and because the dominate emission field of the solar system is the sun's emission field, and because that emission field diminishes inversely with its square from the distance from the sun, the further from the sun an element is positioned, the greater the number of units that element can have in its nucleus and still be stable.

If we use the method of assigning atoms an atomic number, then only hydrogen, with an atomic number of one can exist on the surface of the sun. It is a single unit, the final stage at which all matter breaks down.

On the continuum stretching from the surface of the sun to the orbit of Pluto, the highest atomic number of the atoms that can exist steadily increases until all matter can exist in the orbit of Pluto, from an atom with a single unit in the nucleus to an atom that would have an atomic number of perhaps one hundred and twenty.

THE NUMBER OF UNITS THAT CAN BE BOUND TOGETHER INTO THE NUCLEUS OF AN ATOM IS PROPORTIONAL TO ITS DISTANCE FROM THE SURFACE OF THE SUN, ATOMS WITH THE HIGHEST NUMBER OF UNITS EXISTING IN THE ORBIT OF PLUTO AND ATOMS WITH ONLY ONE UNIT, HYDROGEN, EXISTING ON THE SURFACE OF THE SUN.

We have no way of knowing what the atomic number of an atom with the highest number of units in the nucleus would be. This atom is created in the total absence of a field during the process of atom formation. The number of units would be triple, perhaps quadruple its atomic number, being equivalent to the Earth based perception of weight, the effect of the attractive mechanism on the individual units in the nucleus. However, one hundred and twenty is a possibility.

We do know that the field factual congruity presents us with a continuum of matter with respect to the solar system similar to the continuum of matter we find when we set a fire, with the increase in the field eliminating in an orderly manner "heavier" atoms.

10. THE FIELD FACTUAL CONGRUITY: THE COMET

The comet is modern science's test of faith!

If you believe that a snowball can periodically move close to the sun, emitting vast clouds of water moisture, retreat into the depths of space, only to return to the fiery bosom of the sun to emit still more water vapor, and do this over and over as far as recorded history stretches, then the high priests of the New Religion of Modern Science will know that they can get you to believe any hocus pocus explanation about physical phenomena that they want to make up without danger of exposing their own ignorance about those phenomena.

It takes blind belief to think that an orbiting ice cube could survive the sun, because on Earth the sun melts ice.

But then, space is so cold!

Halley's comet is the best known hunk of ice and snow traveling through space. It regularly reappears, being invisible until it reaches the inner solar system at which point it becomes visible, the head moving in a highly elliptical orbit, with a long tail trailing behind it as it moves toward the sun, the tail preceding it as it moves away from the sun.

During its last visit in nineteen eighty-six, the European Space Agency sent its robot Giotto to get close enough to photograph it.

This hunk of ice turned out to be a rock the size of Manhattan, about ten miles long and five miles on a side, with a one thousand three hundred foot mountain in the middle and a valley at the tip.

This is four times the estimated size.

No Eskimos were photographed, probably because the snow ball had a surface temperature estimated at four hundred on the Celsius scale.

This was one hot rock, which is what might be expected of matter that is approaching an object combusting with the intensity of the sun.

> **IF THE INVERSE SQUARE LAW DICTATES THAT SOMETHING BE MADE OUT OF ICE, AND, WHEN IT'S ACTUALLY MEASURED, IT'S TOO BIG TO BE ICE, THEN IT MUST BE SNOW.**

After all, if we take a paper clip and move it into the field of a burning match, the metal would first become hot, then it would begin to glow, giving off emissions that were in the visible portion of the electromagnetic spectrum.

Why would we assume that the same thing wouldn't happen with matter that was in an orbit that would bring it periodically close to the sun?

But it gets better. When the surface temperature was found to be four hundred degrees centigrade, someone sat down and did some calculating and discovered that ice couldn't survive at that temperature.

It would melt!

And what was worse, it couldn't be ice anyway. If it were four times as large as they thought it was, then it couldn't be ice. It had to be snow, although snow was a little too light, so it had to be dirty snow, snow mixed with dust.

But if ice would melt, snow would surely melt, so the snow had to be in the shade. As there was only one mountain and one valley, the shade had to be in the form of a very thin covering which protected the dirty snow underneath.

IF SNOW DOESN'T MATCH THE INVERSE SQUARE LAW WHEN COMPARED TO SIZE, THEN SCIENCE MERELY ADDS A LITTLE DUST TO MAKE UP THE DIFFERENCE.

It was the breaks in this covering that allowed the water molecules to burst out from the surface from the force of ...what? Well, there must be some sort of gas there too because the particles burst from the surface going at three times the speed of sound, so there certainly had to be gas underneath a thin covering that protects the dirty snow from melting at four hundred degrees Celsius.

And so on and so forth, and so on and on and on and on.

What is it that allows us to make up explanation after explanation, like a kid with both hands in the cookie jar, each one wilder than the last, each one implausible to an impartial viewer, but each one believed more fervently the further from reality it becomes?

What is it that makes us incapable of simply seeing things the way they are, the way facts exist in physical reality?

We will look at that question more closely when we construct the mind out of the elementary particle with its affinity propensity and at-rest motion, but we have already seen why the comet has to be a ball of ice, or snow, or dirty snow, or whatever.

Newton started out assuming that matter generated gravity so that it was proportional to the amount of matter. He did this to prove that the moon was falling toward Earth at the same rate that a rock did on the Earth's surface so that the force of gravity would be universal and inverse with the square of its distance from the source.

AD HOC HYPOTHESES, NO MATTER HOW WILD, ARE ACCEPTABLE SO LONG AS THEY DON'T DISAGREE WITH THE MASS/GRAVITY CONCEPT.

The scientific method has the knack of taking speculations and labeling them as hypothesis when an experiment is devised in which it is thought that the hypothesis can be proven or disproven.

The way the scientific method proves a hypothesis is to let the hypothesis predict a fact, and if the fact is found, then the hypothesis is proven.

We know that hypotheses are merely concepts, and that concepts are only good in so far as they explain facts. A hypothesis that predicts a fact becomes simply a more useful concept, more useful in that it explains an additional fact for which we did not have an explanation.

The scientific method, however, allows us to believe that we have proven a hypothesis if we have found the predictive fact, and the hypothesis then becomes a fact that can be used as a basis for future hypotheses.

When Newton proved that the moon was affected by the same force that affected the rock, and that that force was indeed diminished inversely with the distance over which it acted, he created a fact.

And while I am sure that the moon, were it to fall to the Earth like a rock, would accelerate in a fashion that would prove that the force increased inversely with its distance toward the surface of the Earth, Newton's proof was quickly disproven. Neither the Moon, nor, in fact, anything else had the visual matter to justify the inverse

NEWTON STARTED OUT WITH THE ASSUMPTION THAT THE FORCE OF ATTRACTION WAS PROPORTIONAL TO MATTER, COMPUTED IT ON THE BASIS OF SIZE, AND THEN "PROVED" THE INVERSE SQUARE LAW. WHEN IT TURNED OUT THAT THE INVERSE SQUARE LAW WASN'T PROPORTIONAL TO SIZE, HIS DISCIPLES TURNED IT AROUND, USING THE INVERSE SQUARE LAW TO PROVE HOW MUCH MATTER WAS PRESENT INSIDE THE PLANETS. AND, FOR THAT MATTER, IN EMPTY SPACE.

square law.

However, because the inverse square law had been proven, and because the only forces that existed with respect to planetary movement were the actual movement, an artifact of the origin of the solar system, and the force of gravity, the amount of matter that should be present to cause that movement was easily computed by the inverse square law.

Thus, what Newton set out to prove, that the inverse square law controlled the movement of the planets, now was used to determine what made up the planets.

By reference to how they moved.

Which was known to start with!

If a planet was too big for the inverse square law, it was made out of gas. If it was too small for the inverse square law, it was made out of iron.

We continue to analyze the universe by the limited concepts of weight developed by minds imprisoned on the surface of the Earth.

And if the comets were just a little too light for the inverse square law, then they were ice, and if they were then found out to be a lot lighter than we thought they were because there was more matter there than our calculations had indicated, then that matter must be less dense and must be snow. And because snow melts, there must be a cover to protect it, and because the atoms of matter come off the surface at eighteen hundred miles an hour, then gas must be ejecting them.

And how do we prove all of this to be true.

Well, if water is made out of hydrogen and oxygen, which it is, then we will be able to find a lot of hydrogen on the surface of the comet.

Which we do.

So all of this is now a true fact of the universe, and if our senses tell us otherwise, then our senses are in error, our intuition is faulty, it has to bend in the face of the mathematically proven facts, the realities of modern science.

IF REASON TELLS US A CRUSTED OVER, GAS FILLED, DIRTY SNOWBALL MUST MELT AS IT PASSES CLOSE TO THE SUN, AND IT DOESN'T, IT JUST SHOWS US HOW FLAWED OUR REASONING CAN BE.

Like any great religion, modern science has an answer for every doubt except the ultimate doubt, is the universe really irrational, and that doubt has to be accepted on blind faith, the snowball that repeatedly visits the sun without vaporizing.

What is the comet, if it isn't a snowball?

We have seen that on a continuum that has the sun as its central emitter, Mercury at one end and Pluto at the other, that matter exists in that continuum relative to the strength of the sun's emission field, and that that emission field diminishes inversely with the square of the distance from the sun. If Mercury's orbit were three, and Pluto's three thousand, then the relationship of the field at Mercury's orbit to Pluto's orbit would be nine million to nine.

There would be very few atoms that would hold together on Mercury's surface, and very few that wouldn't on Pluto's.

If Halley's Comet were made up of atoms with an average atomic number of one hundred and fifteen, then when the comet was positioned in the orbit of Pluto, the internal atomic structure of the matter would be in equilibrium. The affinity propensities of the sun's field would balance the internal affinity propensities of the comet's atoms so that the comet would exist as stable matter.

It would not be emitting elementary particles because there would be no field replacement going on.

We know that the comet is not going to stay in the orbit of Pluto because we know that the field that it is in, the expanding sphere of the sun's emissions contains a mechanism that allows it to expand uniformly over the surface of the expanding sphere by alternately causing the expanding flows to recombine.

This mechanism, through its recombination of the expanding flows, acts on the individual units in the nuclei of the atoms of the comet to cause the comet to move toward the source of emissions, the sun.

The source of the matter that makes up the comet itself

is of no concern. The sun may be traveling through space at about one hundred and forty miles a second, which means that it is covering about ten million miles in a day.

The space that it is traveling though has a field that is the total of all the expanding spheres of star light that exist at any point, but that total field is extremely weak, so that atoms with high atomic numbers could easily exist, and the matter that made them up would not have undergone combustion when stars ignited to make up the galaxy the sun occupies.

COMETS ARE MADE UP OF ATOMS WITH A HIGH NUMBER OF UNITS IN THEIR NUCLEI, ATOMS THAT BECOME INCREASINGLY UNSTABLE AS THEY MOVE DEEPER INTO THE SUN'S EMISSION FIELD AND WHICH BECOME EMITTERS, TINY SUNS, AS A RESULT OF FIELD REPLACEMENT.

At ten million miles a day, eventually the sun's expanding sphere of electromagnetic emissions is going to pick up rocks that contain matter that have atomic weights anywhere up to the highest number that can exist in the absence of a field.

It is not important where the matter that makes up comets comes from, only what the matter that makes them up is.

As the matter that makes up Halley's comet leaves the orbit of Pluto and begins to be drawn into the sun's emission field, field replacement is going to begin to occur. The directed field of the sun's emissions, being stationary with respect to the atoms that make up the comet, will begin to replace the affinity propensities of the orbiting elementary particles, and they will begin to be released.

Then, it's just a matter of moving a paper clip closer to the flame, except the atoms of the comet are much more volatile than the atoms of the paper clip, as the phosphorous of a match head is more volatile than copper.

The comet moves past the orbit of Jupiter into a much stronger field. As it gets deeper in the sun's emission field, the elementary particles no longer needed to maintain internal equilibrium start to leave the structure, and their leaving is measurable in the form of heat.

WHAT GIOTTO SAW IS WHAT EXISTS, A HOT HUNK OF ROCK.

As the comet moves closer to the sun, the heat increases until there are too many elementary particles to be emitted over the surface available for emission. They start lining up and structuring themselves into the electromagnetic spectrum, which, as the comet moves deeper into the emission field, itself becomes an emission field, visible in the Earth's sky.

What we expect to happen on the surface of matter undergoing combustion happens on the surface of the comet. The combusting matter breaks down into its simplest form, a single unit of matter.

It is measurable as a hydrogen atom.

The combination of snow and Newtonian density dissolves in a flash of departing elementary particles as the comet undergoes combustion at an increasing rate as it moves deeper into the sun's emission field.

The rock that is reconstructible from Giotto's pictures is just that, a rock, but a rock composed of atoms that can't exist in a field as strong as the sun's field in the inner solar system.

When the comet orbits the sun, and begins its movement away, back toward the orbit of Pluto, it might pass close to not only the orbit of Jupiter, but if the timing is right, it might pass by Jupiter itself.

THE LAST TIME AROUND, HALLEY'S COMET HAD A BRIEF SPURT OF ACTIVITY FOR A SNOWBALL. IT UNEXPECTEDLY BECAME AN EMITTER ONCE AGAIN.

By the time it gets in Jupiter's neighborhood, the sun's emission field would be so weak that the comet would have regained its equilibrium, or at least if it were still a ten mile long hunk of hot rock, it would be well on its way to having its internal affinity propensities re-balanced.

It would have ceased to emit light in the visible spectrum.

Jupiter exists in the sun's emission field so that the matter that makes up Jupiter is balanced in the context of the strength of that field.

We know that matter exists relative to the field in which it finds itself, and that the field of Jupiter is a combination of the field generated by the cooling matter in the planet as that matter undergoes the combustion process, and the sun's emission field which at this point is very weak.

If the sun should have an unexpected storm, what would its effect be on the comet?

On the sun, a storm is considered to be a series of solar flares that erupt as the elementary particles building up for release at the surface of the sun surpass the ability of the sun to keep them waiting. The resulting emission drags the actual surface of the sun out deep into space in a flare. The result of the flare is an increase in the overall emission field of the sun.

The sun is, for the moment, hotter.

These solar flares alter the rotation of the Earth, causing it to slow down for their duration. Because the portion of the Earth that is undergoing combustion, the area some hundred miles beneath our feet, exists with respect to the increased field of the sun, its rate of combustion also increases for the duration of the solar flares, just as the paper clip would get hotter if the burning match hit a second spot of sulfur and flared up, the flame moving closer to the metal of the paper clip.

Heat makes things hotter!

As the Earth heats up momentarily, the sphere of

electromagnetic emissions becomes stronger. Because these emissions have presence, and the presence is relative to the distance from the source of the emissions, increasing the emission field, raising the temperature of the emitter, will increase that presence at any point in the expanding field.

As the attractive mechanism that operates within this field is relative to the presence of the field, an increase in the field will result in an increase in the force of the attractive mechanism.

As solar flares heat up the Earth, causing an increase in the Earth's emissions, the attractive mechanism also increases, causing artificial satellites to be tugged from their orbits.

THE SUN AND PLANETS ARE LOCKED INTO A DYNAMIC SYSTEM IN WHICH A CHANGE IN THE RATE OF COMBUSTION OF EACH EFFECTS THE TEMPERATURE AND MOTION OF ALL.

But what about Halley's comet moving out, deep in the dark sky, passing near Jupiter?

The sun's emission field, increased by the solar flare activity, is too weak to turn the comet into an emitter, it being affected by the emission field of the planet it is close to, Jupiter.

However weak it is, the sun's increased emission field still effects the planet, causing it to heat up.

As the emissions from Jupiter increase, the emission field starts to replace the affinity propensities of the orbiting elementary particles that are working to regain equilibrium within the matter that makes up Halley's comet.

The comet spurts into life momentarily, reflecting the unexpected field replacement caused by the solar activity on the surface of the sun.

11. THE MOVEMENT OF MATTER IN SPACE: ATTRACTION

Newton's clockwork solar system put the planets in motion by a historical force, and therefore removed from consideration the forces that move the planets.

Any objective analysis would conclude that the concept of gravity as a straight line force does not result in circular motion.

When present day scientists construct the beginning of the solar system in their mind, they work back from the fact that everything is traveling in a circle. The planets rotate in the same direction, they orbit in the same direction, the moons orbit the planets in the same direction, the sun rotates in the same direction, the matter that makes up the asteroid belt orbits in the same direction.

SWIRLING MASSES OF GAS CAN BE ADDED TO MASS/GRAVITY AND LIGHT AS WATER WAVES TO CREATE THE TRINITY OF IGNORANCE.

And all of this happens in the equatorial plane of the sun.

Under a sort of modification of the "I think therefore I am" proposition, modern day scientists think the solar system is as it is, therefore it must have gotten that way by condensing out of a swirling mass of gas.

Even if gravity was a property of mass, and the dust particles that made up the mass of gas were each imbued with their allocation of mass, each attracting every other in the mass, they would all end up in a clump of matter.

There is no force to set them swirling!

And, that clump of matter would always have the greatest attraction on its surface because the opposing attractions of the particles in the sphere cancel each other out until weight does not exist at the center.

Thus, there would never be the pressure necessary for the matter to explode in a big bang that would conveniently get everything started in a universe that exists only in the mind of modern science.

No. I think rather, given the mass gravity concept, all matter would eventually end up in one big ball floating endlessly in space, with no size because there would be no other clump of matter to compare it with, and no movement, for the same reason.

Pretty hopeless.

As hopeless, in fact, as the universe that has been created by the mass/gravity concept and then mindlessly twisted into unreality by assumptions that defy reason.

If you ask the scientist thinking about the swirling mass of gas where the swirl came from, he will tell you that there are electromagnetic forces that are not understood with clarity, but which, he can assure you, exist just as surely as the planets are moving in their mathematically predictable orbits around the sun.

The basis of the swirling mass of gas nonsense is the

Newtonian proposition that matter will move in a straight line unless a force acts upon it to make it move in other than a straight line.

Newton needed this assumption to calculate the momentum of the moon, which he in turn needed to calculate the force that was overcoming that momentum.

Gravity was a force that existed, and was always there.

The moon was always moving.

Whence the momentum came really didn't matter, just as it didn't matter what was causing the gravity.

It existed.

It was measurable.

That was all that was needed to know.

The key took another successful turn in the ignition!

NEWTON WAS NOT CONCERNED WITH HOW MATTER PRODUCES GRAVITY, JUST AS MODERN SCIENCE IS NOT CONCERNED WITH WHAT MAKES MATTER MOVE.

The operation of physical phenomena was, and is, according to the key concept, just a simple matter of coming up with the right mathematical system to allow the phenomena to be measured.

The hell with reason.

And that is exactly what Newton was saying when he said that an object will move in a straight line unless a force acts on it to make it move in other than a straight line.

However, objects don't move in straight lines.

Objects come to rest with respect to the forces acting upon them.

This is a horrible wrench to a mind that is used to having all the motion exist in the solar system without reference to existing forces.

If you place the cause of all the motion in the solar system five billion years in the past, you don't have to worry about coming up with an explanation for it.

If you don't have to explain something, you won't find yourself embarrassed when you can't explain it.

And who would want to admit in this day and age when all but the very last secrets of the universe are just another Bubble telescope or jumbo super-duper collider away that they don't know something as obvious as why the planet rotates on its axis.

Or why the sun comes up in the morning!

Nobody, if they wanted to keep their hand in the public till.

But, unfortunately, matter doesn't move in a straight line until a force acts on it to move it in another way.

Matter comes to rest with respect to the forces that are currently acting upon it.

Present forces, not historical forces.

Forces that exist!

Forces that require an explanation if we are going to come up with a consistent picture of physical reality.

This involves a complete reversal in perspective.

Instead of the moon rumbling along in its orbit, moving around the Earth about every twenty-nine days with reference to the sun, or twenty-seven with reference to the stars, the moon is dead still.

It is at rest!

At rest with what?

At rest with respect to the forces that are currently acting upon it.

With respect to the forces currently acting upon it?

THE PLANETS DO NOT MOVE THROUGH SPACE AS A RESULT OF HISTORICAL FORCES THAT ARE NO LONGER AROUND. THEY MOVE AS A RESULT OF EXISTING FORCES.

What forces?

That's the question.

Because not only is the moon moving in orbit around the Earth, the Earth is moving around the sun, clocking about 67,000 miles per hour. The Earth is slow when compared to Mercury, which is moving at over a hundred thousand miles an hour. As you move out in the continuum of the sun's emission field, the speeds of the planets decrease until you get to Pluto, which is only moving at about eleven thousand miles per hour.

Many of these planets have moons that travel around them.

And of course the sun, a star that is in one of the arms of the galaxy viewed as the milky way, is moving around the center of that galaxy at about one hundred and forty miles a second, or about half a million miles an hour, and in the process, dragging the planets behind it in a fashion that causes their orbits to trail away from the direction of motion only to have to play catch up, resulting in elliptical rather than the circular orbits that would be expected from a clockwork system.

A SOLAR SYSTEM OPERATING WITH CURRENT FORCES IS DYNAMIC, FLEXIBLE. WITH ITS MATTER MOVING TO REMAIN IN BALANCE WITH EXISTING FORCES, IT EASILY SURVIVES CHANGES IN THOSE CURRENT FORCES. A STATIC SOLAR SYSTEM OPERATING ON THE BASIS OF HISTORICAL FORCES WOULD QUICKLY PASS INTO HISTORY.

If a non-melting snow ball doesn't provide a test of faith, then surely the idea that this complex system in perpetual movement does not have current forces holding it together surely will!

But then, all you have to do is look at the moon speeding along, and the idea that these planets are motionless is perhaps a greater test of faith.

From a solar system that seems dynamic, with the

matter that makes it up in perpetual motion, to a perfectly balanced solar system, in which the matter is in motion with respect to other matter, but motionless with respect to the forces creating the motion, this takes a conceptual leap.

And yet, the concept of the solar system as one put in perpetual motion, with unchanging orbits and inflexible rotation, is a static concept, rather than a dynamic concept.

A solar system operated by historical forces, in which the matter was put in motion five billion years ago, and which therefore is fixed, is a static solar system.

The dynamic solar system would be one where the movement of matter was governed by existing forces, and adjusted itself to accommodate the changes in those existing forces.

How much force is required to move matter?

In the absence of force, the mere presence of force will move matter without regard to the physical size of the matter, or the number of units that make up the atoms of the matter.

If we go back to our laboratory of non-existence, we can flood the area in which there is no existing field with elementary particles.

Their at-rest motion will eventually be overcome by collisions that will cause units of matter to form, and in turn form together into the nuclei of the most complex atom that can form.

These atoms in turn will move together into

equilibrium so that eventually the matter, which because it is forming on all available areas, is forming into a sphere, takes shape and reaches the size of the sun.

The matter has not yet ignited.

What would it take to move that matter in space?

Because there is no force affecting the units of the atoms that make up the physical structure, a feather stroke would set all the atoms in the structure in motion at the same time.

THE MEREST FORCE WILL AFFECT THE MOTION OF MATTER WHERE NO FORCES ARE AFFECTING THE MATTER, OR THE FORCES THAT ARE AFFECTING THE MATTER ARE IN BALANCE.

The matter is held together by its internal forces, which are balanced. There is no force affecting the units of the atoms. Thus, pressure on one side of the sphere would affect the other side of the sphere.

There is no inertia to overcome because there is no force acting on the individual units of the atoms to provide the inertia.

The mere touch of a feather would set the structure in motion.

Once the structure is in motion, it would take the same touch of the feather to stop it from moving.

The same situation holds true for the matter that makes up the solar system.

If the matter is at rest, in equilibrium with the forces in the environment in which it exists, then it takes the merest change in the force to effect the equilibrium, and that change would effect the motion of the matter that is at rest with respect to those forces.

The concept is viewable in the television demonstrations of weightlessness. The object floating in front of the astronaut is in equilibrium with respect to the forces acting on it, in this case the Newtonian concept of a projectile's forward motion balancing the force of gravity.

The astronaut can touch the object with a feather. The feather would not move the object on Earth, but in the space capsule, the object responds easily.

When there is no force acting on matter, or where the forces acting on matter are in equilibrium, then it takes the merest existence of force to affect the motion of that matter.

The space capsule is in the same equilibrium as the object floating within it.

However, from the Earth, the capsule is moving at a rapid rate across the night sky.

In the capsule, the object is also moving at a rapid rate across the night sky. However, with respect to the closed system of the capsule that we see on the television screen, it is stationary.

And indeed it is stationary, just as the capsule is stationary.

It is stationary with respect to the inside walls of the

capsule, but it is also stationary with respect to the invisible forces that are acting upon it.

If we bring the object and the capsule back to Earth, it regains its weight. It now takes more force than that of the mere presence of force to move it, because the Earth is generating an expanding field of electromagnetic emissions that produce a mechanism that causes the units of the atoms of the matter to move toward the source of emissions.

However, the Earth exists with respect to the sun the same as the capsule exists with respect to the Earth.

The Earth is weightless, just like the object in the space capsule is weightless, and the mere presence of force will effect the Earth's movement.

BECAUSE THE EARTH EXISTS IN EQUILIBRIUM WITH THE FORCES THAT ARE ACTING ON IT, THE MEREST CHANGE IN THOSE FORCES WILL CHANGE THE MOVEMENT OF THE EARTH.

The difference between matter existing free of force, and matter embedded in a field of force is that when matter is embedded in a field of force all the units in all the atoms that are affected by the force have to be moved uniformly.

This fortunately prevents someone falling off a stool from affecting the orbit of the planet, the force of the fall affecting only a small area of the planet's surface.

However, when the force is embedded in the field of sunlight, and the field is uniformly distributed over the surface of the planet, the merest change in the force embedded in that field affects the movement of the planet within that field.

THE RELATION BETWEEN SOLAR ACTIVITY AND DECAY IN THE ORBITS OF ARTIFICIAL SATELLITES IS CLEAR. THE CAUSE, HOWEVER, IS NOT COMPLEX CHANGES IN THE SOLAR WIND RESULTING FROM SUN SPOTS. THE SUN'S EMISSION FIELD INCREASES, HEATING UP THE EARTH. THE EARTH'S EMISSION FIELD INCREASES, INCREASING THE FORCE OF THE ATTRACTIVE MECHANISM EMBEDDED IN ITS PROCESS OF EXPANSION. THIS INCREASE CAUSES THE ARTIFICIAL SATELLITE TO FALL TOWARD THE EARTH AT A GREATER RATE, DECAYING ITS ORBIT.

And the Earth does change its rate of rotation, imperceptibly perhaps, but it's not the weather that is causing this change in rate any more than it is gravity that is causing the planets to rotate in the first place.

Changes in the rate of rotation are associated with solar activity, just as decay in the orbits of artificial satellites is associated with solar activity.

To see how the movement of matter can develop, lets move back into our laboratory of non-existence, and create a second sphere of matter the same size as the first so that we have two spheres of matter the size of the sun.

If we put the second sphere of matter as far away from

the first as Pluto is from the sun, and light a match next to its side, we won't have to wait long before the expanding sphere of light from the match starts to replace the affinity propensities of the orbiting elementary particles in the atoms at the surface of the sphere.

As the elementary particles begin to depart the surface at their at-rest speed, the streams in turn act like directed flows and replace additional affinity propensities which in turn increase the flows which in turn create more directed streams, and so forth.

Pretty soon, a patch on the surface of the sphere has ignited and is emitting elementary particles in the form of visible light.

While some elementary particles, when they are released from their orbits, shoot into the interior of the sphere, replacing additional elementary particles as they pass through the structure and out the other side, they do pass out the other side, adding to the elementary particles that are being emitted at the surface of the sphere.

At the same time, the patch of emission on the surface where the match started the process is spreading in all directions out over the surface of the sphere.

As the star flares into life, the elementary particles begin to line up at the surface for emission, and the excess affinity propensities created by elementary particles that have regained their at-rest motion but can't go anywhere keep additional particles from moving into the center of the structure.

In any event, the center of the structure does not begin to combust simply because there is no where for the elementary particles that would be emitted in combustion to go.

Combustion takes place on the surface because the surface is the area available for the emission of elementary particles that are regaining their at-rest motion.

As soon as the streams of particles are emitted in the expanding sphere of electromagnetic radiation, inductive flows swirl around them. These inductive flows are registered as electromagnetic by detectors designed to measure the inductive flows.

It is, perhaps, these electromagnetic measurements that give rise to statements that planetary and galactic motions are governed by electromagnetic forces which lack clarity at present and which therefore are not worth explaining, being in the realm of philosophy.

The inductive flows produce the attractive mechanism that has the dual purpose of regulating the expansion of the light being emitted.

Soon, the entire surface of the second ball of matter is a star ablaze, and is producing an expanding sphere of electromagnetic emission.

As the sphere expands, it encompasses the first sphere of matter, and, acting like our match, ignites that sphere that is also soon ablaze.

We have two stars sitting several billion miles apart,

both emitting expanding spheres of electromagnetic emissions.

What is going to happen?

Each star is going to be embedded in the other star's expanding emission field. Each of those fields is expanding over the surface of the expanding sphere, and as it does so, it is producing at each point in the expanding field a mechanism that is both regulating the expansion of the field and creating a force back towards the source of the field that can act on the units in the nuclei of any atoms that become imbedded in the expanding field.

As both stars are embedded in each other's field, they both start to move in the direction of the attractive mechanism.

They both start moving toward each other!

If gravity were the only force at play, then it is obvious that the two stars would crash into each other, possibly explode, shattering themselves into tiny stars that would proceed to attract each other with the same result, the process going on until everything had been reduced to elementary particles moving at their at-rest speed.

Fortunately, gravity doesn't exist, at least not as a property of matter, and the attractive force, which is caused by the interplay between the expansion of the emission field over the surface of the expanding sphere and the inductive flows created by the emissions presence at any point in the emission field, is not the only force in operation.

12. THE MOVEMENT OF MATTER IN SPACE: ROTATION

The second force in operation is the force produced by the direction of the inductive flows that orbit the stationary presence of the affinity propensities of the emission fields.

This flow, as in electricity, operates on the basis of the right hand rule. If you put the thumb of your right hand in the direction of the flow, the fingers curve in the direction of the inductive flow.

THE ATTRACTIVE MECHANISM EMBEDDED IN THE EXPANDING ELECTROMAGNETIC SPECTRUM MOVES OBJECTS TOWARD COMBUSTING MATTER, CAUSING IT IN TURN TO IGNITE AND COMBUST. THE INDUCTIVE FLOWS OPERATING AT RIGHT ANGLES TO THE FLOWS ESTABLISH SPIN IN THE OBJECTS.

If one of our stars is on the left, and the other one is on the right, then the emissions are flowing toward a point in front of us.

If we align the thumb of the right hand with the flows from the star on the right, we see that the inductive flows curl in an orbital movement that is moving away from us at its top arc.

If we align the thumb of the right hand with the flows from the star on the left, we see that the inductive flows curl in an orbital movement that is moving towards us at its top arc.

The inductive flows between the two emitters move in opposition to each other.

What would result from this situation, if there were not an attractive mechanism associated with the expanding emission field would be two stars that would begin to rotate as the inductive flows opposed the inductive flows on the surface of the opposite stars.

The direction of rotation is a simple problem of mechanics.

If you attached two sticks to the surface of the stars so that they were parallel with each other between the stars, and then you moved the sticks apart, the stars would rotate, and the rotation would be in the same direction for both stars.

THERE ARE ONLY TWO ASYMMETRIES IN THE UNIVERSE: THE RIGHT HAND RULE AND COOLING. IT IS REASONABLE TO ASSUME THAT ONE RESULTS FROM THE OTHER.

Thus, the opposing inductances established by the presences at each point in the emission flows will establish rotation in emitters embedded in the emission fields, and the rotation will be in the same direction.

How do we determine direction in a space filled with nothing?

Direction is determined by the fact that combusting matter cools, and in doing so establishes the direction of the inductive flow.

When the opposing inductive flows between emitters establish rotation in the emitters, the matter that is rotating acts as if it itself were an inductive flow, the elementary particles of the matter orbiting the axis of rotation.

Thus, the direction of rotation establishes the direction of the flows through the matter, the flows being established by the direction of the rotating matter mimicking an inductive flow.

IF WE ASSIGN NORTH AS THE PLACE TO WHICH ELEMENTARY PARTICLES FLOW, THEN THE FINGERS OF THE RIGHT HAND CURL IN THE DIRECTION OF PLANETARY ROTATION WHEN THE THUMB IS POINTING TO THE NORTH POLE.

If we put our fingers in the direction of the rotation of the Earth, then the thumb points in the direction of this flow, which is to the north pole.

The flow, made up of elementary particles, exits the north pole, and acts like what are measured as magnetic lines of force, moving out and around the planet, making an intersection with, and joining new particles induced into the south pole of the planet. It is these flows that line up with the external flows of magnetic matter, causing the matter itself to line up north-south. As north is defined as the direction to which the elementary particles flow, then the magnet, by aligning its external flows with the Earth's external flows, points its north end north, giving the lie to the recent realization that because opposites attract, it must be its south end pointing toward the north pole.

The direction at which a planet rotates is determined by the right hand rule applied to the opposing inductive flows of the emissions between the planet and the dominate emitter in the system, the sun.

Without the attractive mechanism in the emission fields, the opposing inductances between the two stars would establish rotation in the stars and the rotation would be in identical directions with respect to both stars. This direction is determined by pointing the thumb of the right hand up through the axis of rotation.

CENTRIFUGAL FORCES CONCENTRATE THE EMISSION FIELD AT THE EQUATOR.

The fingers will point in the direction of the rotation.

We can determine the direction of orbiting the same way, but for different reasons.

We can easily see the configuration of the emission field of a rotating object by looking at how stars arrange themselves in that field.

We can do this by looking at a galaxy.

The galaxy's shape is defined by the stars that make it up. As these stars are embedded in each other's emission fields, the location of the stars defines the nature of that emission field.

The position of the stars that define the emission fields of the rotating emitters that make up the galaxy, the stars, indicates that the emission fields of rotating emitters do two things as a result of rotation.

First, because the emitters are made up of matter that is losing its structural cohesiveness, rotation will produce centrifugal forces that will cause that matter to move toward the equator as established by the rotation.

Masses of combusting matter, the hottest, and the most concentrated will migrate to the equator of the sun.

This will result in an emission field that is strongest at the equator of the sun.

AS THE ATTRACTIVE MECHANISM IS EMBEDDED IN THE EMISSION FIELD, THE STRONGEST AREA OF ATTRACTION WILL BE IN THE EQUATORIAL EMISSION FIELD, GIVING THE SOLAR SYSTEM (AND GALAXIES) PLANULARITY.

As the attractive mechanism is proportional to the emission field, the strongest attractive force in the expanding sphere of electromagnetic emissions will be in the equatorial region of the star, the region in which the stars of a galaxy are located, and the region in which the planets orbit the sun, and the moons orbit the planets.

The next thing we can learn about the emission field that results from the rotation of the emitter is that the field trails the direction of rotation.

The stars embedded in the emission fields of the stars that make up the galaxies outline emission fields that are emitted, but being coherent flows of elementary particles, move out from the sphere of the emitter, and in doing so, create flows that fan out away from the direction of rotation.

The effect is seen clearly in the arms of a spiral galaxy, which trail out behind the center. If the stars are used to trace the forces in which they are embedded, then it is clear that the emission fields of the stars trail out and behind the rotating center.

ROTATION CREATES A REVERSE BIAS IN THE ATTRACTIVE MECHANISM, CAUSING MATTER WHICH IS CAUGHT IN ITS FORCE TO FALL IN THE DIRECTION OF ROTATION RATHER THAN DIRECTLY TOWARD THE EMITTER.

The reason for this is physical.

The surface of an emitter moves forward the diameter of an elementary particle by the time the next group of elementary particles leave its surface. By the time the second particle of the stream is emitted, it is being emitted ahead of the flow of the particle that was emitted the instant before.

Therefore, the streams are always trailing the direction of flow, fanning out in what appears to be a backward direction in an arc away from the direction of rotation but what is really the physical result of having been emitted from a rotating emitter.

The expanding field of electromagnetic emissions, the light from a rotating star, therefore has a reverse bias away from the direction of rotation.

Light doesn't move from the surface of the sun to the orbit of Pluto in a straight line.

The expanding streams of light that reach Pluto describe an arc from the surface of the sun to Pluto, with the curvature of the arc being in the direction that Pluto is moving.

As the expanding streams of light contain the mechanism that regulates its expansion, as well as produce the force of attraction back in the direction of expansion, the attractive force does not move units of atoms back directly toward the surface of the sun, but rather back along the arc of expansion as established by the rotation of the sun.

THIS REVERSE BIAS MEANS THAT MATTER GAINS MOMENTUM AS IT FALLS, THE RESULTING ORBIT BEING A BALANCE BETWEEN THE MOMENTUM AND THE RATE OF FALL.

Pluto will therefore tend to fall in an arc toward the sun, rather than directly toward the sun.

And here, amazingly enough, comes clear the concept of an orbiting planet accelerating to maintain its orbit, a concept that is totally ridiculous with gravity as the sole current force and momentum the result of a historical force.

The planets are being attracted, and accelerated inversely with the square of the distance they move along the arc over which the emission field is expanding.

As a planet accelerates, it gains momentum without reference to the affinity propensities holding the units of the atoms that make it up together.

However, the momentum works against the attractive mechanisms in the expanding field, and the affinity propensities that hold the units of the atoms together give the planet what we perceive as weight.

ALL MATTER IN THE SOLAR SYSTEM MAINTAINS STABILITY BY READJUSTING ITS POSITION IN RESPONSE TO CHANGES IN THE RATE OF COMBUSTION OF EACH.

This weight, which is now momentum, balances the rate of acceleration to stabilize the planet's orbit.

And the operative word is stabilize.

It does not fix the orbit in place.

An increase in solar activity would increase the force of the attractive mechanism enough to imperceptibly alter the acceleration at which the planet is falling. The increase in acceleration would increase the momentum which in turn would alter the orbit slightly.

When the solar activity ceased, the planet would readjust to the new rate of cooling, the new emission field.

What is the rate of acceleration for Pluto as compared with, say the Earth?

As we can all remember from watching the antics of the astronauts on the moon, a lead ball and a feather will drop in an emission field at the same rate, the emission field of the moon being the measured, but non-believed fact that the moon is still generating heat.

As the objects fell, they fell at the same rate compared with each other, but they fell at a different rate than they would have fallen had they been dropped in a vacuum on Earth.

While the rate of fall of an object in an expanding field of electromagnetic emissions will always be measurable as a product of the inverse square law, distances that produce the inverse measurement will increase the stronger into the field the acceleration occurs, or the stronger the field is.

WHILE MATTER FALLS AT A RATE THAT ACCELERATES INVERSELY WITH THE SQUARE OF THE DISTANCE OVER WHICH IT FALLS, THE DISTANCE OVER WHICH IT ACCELERATES INCREASES PER UNIT OF TIME WITH THE STRENGTH OF THE EMISSION FIELD.

Just as the solar activity would increase the force of the acceleration of Pluto, increasing the distance over which it accelerated, so would being deeper in the field increase the acceleration.

The Earth is falling faster than Pluto because the Earth is closer to the source of emissions and is therefore in a stronger field. The attractive mechanism is proportional to the presence of the field. The closer to the sun, the greater is the force of attraction and the greater the rate of fall.

If we go back to our two stars, existing in non-existence, and put the attractive mechanism back into the expanding emission fields, we have two stars, each embedded in the other's expanding field, rotating because of the opposing inductances of the flows, and being attracted to each other, not in a direct line between each other, but over the arc of the expanding flows, the arc created by the star's rotation.

THE REVERSE BIAS OF THE ATTRACTIVE MECHANISM PREVENTS COMBUSTING OBJECTS FROM COLLIDING, SOMETHING THAT WOULD BE UNAVOIDABLE IF GRAVITY WERE THE STRAIGHT LINE RESULT OF MASS.

The two stars will start to move toward each other, not directly toward each other, but close enough.

What happens as they move toward each other?

They each become embedded deeper in the other's emission field.

How does this affect the stars?

First, the rate of combustion, field replacement, is proportional to the strength of the field in which the matter undergoing combustion finds itself.

As the stars move closer, they are going to burn brighter.

As they burn brighter, their emission fields are going to

increase.

As the emission fields increase, both the attractive mechanism, which is moving them towards each other, and the inductive flows, which are causing them to rotate, are increasing.

As the rotation of the stars increases, the bias of the emission fields against the direction of rotation increases.

As the bias increases, the angle of the movement away from each other increases.

If we move our perspective so that we are over the two north poles of the two stars, then we see that they begin to move toward each other as if they might pass each other on their left with respect to the direction of travel. As they start to heat up, as evidenced by their color moving from the red end of the spectrum to the blue end of the spectrum, we notice that they begin to speed up, their rate of rotation increases, and they begin to look like they are widening the distance that will be between them when the pass each other.

We might watch in fascination as the closer they get, the faster they move, the hotter they get, the faster they rotate, the bigger the angle away from impact becomes.

We might realize that it is impossible for two combusting bodies in space to collide because the attractive mechanism causing them to move together, the field causing them to undergo combustion and the inductive forces causing them to rotate are all proportional to the same thing, the presence of the emission field from the combustion, and indeed, are the same thing, the interaction of the elementary particles that make up that emission field, the elementary particles with their properties of at-rest motion and affinity propensity interacting.

Rather than collide, the two stars move off at right angles with one another, their momentum carrying them away from each other.

THE REVERSE BIAS OF THE ATTRACTIVE MECHANISM CAUSING MATTER TO FALL IN THE DIRECTION OF ROTATION IS CLEARLY REFLECTED IN THE BINARY STAR SYSTEMS WHICH COULD NOT EXIST WITH GRAVITY AS A STRAIGHT LINE RESULT OF MASS.

In the process the three physical results, acceleration, rotation and temperature, diminish.

When the attractive mechanism overcomes this momentum, the two stars are already traveling in a counterclockwise motion with respect to one another. The next pass will result in an even shallower passage, with the two hardly coming close to one another as their increasing momentum carries them once again away from each other.

As the process continues, eventually they reach a point with respect to one another in which their rate of combustion, their momentum, their acceleration and their rotation are all in balance, in equilibrium.

Once the equilibrium is reached, if there are no outside

influences, the two stars would orbit each other in a boring fashion, as if some sort of force had put them in motion an undetermined time before.

Of course, their orbits would be driven by their rate of combustion. Because the stars were cooling, the orbits would be decaying on an imperceptible basis. Their temperature, rotation, acceleration, and momentum might all increase imperceptibly as a result of solar activity that results in the build up of more elementary particles than are willing to wait for their orderly emission in the electromagnetic spectrum, the solar activity being an actual eruption of the surface in a solar flare.

Or the solar activity might evidence itself in sun spots, the result of inductive flows of each sun's emissions, and on occasion, the opposing inductive flows on the surface of each sun.

If a third hunk of matter the size of the two suns happened along, the resulting three combusting bodies of matter would mix it up once again, this time with the same impossibility of the three colliding, and the same result of orbital equilibrium, with the established orbits being ones that put each of the stars in equilibrium with respect to temperature, acceleration, momentum and rotation.

From the north polar viewpoint, the rotation would always be counterclockwise, the orbiting would be counterclockwise, and would occur in an equatorial plane established by the rotation of the three stars.

If the three stars could not accommodate one another,

could not come to rest with respect to the forces acting on them, the forces contained in the emission fields, the forces of attraction, induction and presence, then one would be forced out of the system.

Matter just does not move in space in repetitive patterns unless it is in equilibrium, at rest with the forces that affect it.

RATHER THAN CONDENSING OUT OF A SWIRLING MASS OF GAS, THE SUN AND PLANETS OF THE SOLAR SYSTEM FORMED ATOM BY ATOM IN THE ABSENCE OF A FIELD. WHEN IGNITION OCCURRED, THE MATTER SORTED ITSELF OUT BY THE RATE AT WHICH EACH WAS UNDERGOING COMBUSTION, AND SIZE, MOVING EVENTUALLY INTO BALANCED ORBITS IN THE EQUATORIAL PLANE OF THE DOMINANT EMITTER, THE SUN.

If we move back into reality, we can envision an area of space the size of a future galaxy. Matter formation has been going on for some time, and in fact the matter in the center of the galaxy has already started to ignite, the expanding fields encompassing other spheres of matter causing them to ignite in turn.

The new stars at the center of the field of matter formation are just beginning to establish their positions within the structure, but the emission fields have not yet reached the edges of the overall sphere that makes up the area defined by the lack of a directed emission field in which the matter formation occurred.

In our little area of the sphere, matter formation had begun, creating a large sphere of matter the size of the sun, four smaller ones, the size of the outer planets, and to avoid argument, four even smaller ones, the size of the inner planets.

This was the situation when the starlight from the center of the galaxy reached the area of our future sun for the first time.

The directed fields that made up the starlight ignited all the matter at once. The sun, and each of the nine planets all of a sudden became emitters, existing in one another's fields, affecting each other's temperature, rotation, acceleration and momentum.

They all have to sort themselves out.

It is easy to see that the smaller emitters will end up accelerating in the emission field of the dominate emitter, the sun.

It is easy to see that the orbiting will occur in the direction of the dominate emitter's emission field, the direction of fall established by the arc of the emissions created by the rotation.

It is easy to see that this rotation will be the same in all the emitters.

It is easy to see that rotation of the dominate emitter will create an equatorial plane in which the orbiting will occur.

It is easy to see that the large planets would have to

take up positions further away from the dominate emitter than the smaller planets because the momentum would be controlled by the number of atoms bound together, their weight. The momentum would increase with acceleration, which would increase the closer the planet was to the emitter. The greater the momentum, the further away from the dominate emitter it would have to be for it to achieve equilibrium.

It is easy to see why the speed of the planets diminishes the further from the dominate emitter, the sun, that the equilibrium places them, with small planets like the Earth whizzing around at almost seventy thousand miles an hour while Jupiter is lumbering along at less than half the speed, and Uranus at half again the speed, with Pluto skipping around at a mere eleven thousand miles an hour.

IT IS MORE COMFORTING TO LIVE IN A DYNAMIC SOLAR SYSTEM THAT CAN ADJUST ITSELF TO CHANGE THAN IT IS TO BE SAFELY STUCK ON THE SURFACE OF A PLANET WHOSE COURSE WAS SET EONS IN THE PAST AND WILL CONTINUE TO MOVE COME HELL OR PLANETOID.

In fact it is easy to see how the solar system came to be as it is.

And it is quite a relief to know, that should the solar system, as it travels its twelve million miles a day through space, pass an unignited hunk of matter as big as a planet and ignites it, pulling it into the solar system, it won't be the end of the solar system.

The solar system will just achieve a new equilibrium if it can accommodate it, and the planets will resume their steady movement around the sun.

A movement made steady because combusting matter cools, it does so imperceptibly, and in a straight line from hot to cold.

The following Chapter Thirteen introduces the concept of matter as physical, chemical, or equilbrical.

The first twelve chapters dealt with physical matter.

Chapter Fourteen deals with the development of chemical matter, life, as a result of physical matter being subjected to a continually changing field.

Chemical matter is not a concept.

Chapter Fifteen introduces the concept of equilibrical matter. The extent to which equilibrical matter is considered a concept depends on how real the mind is considered to be.

Equilibrical matter is simply a structure formed from the elementary particle with its at-rest motion and affinity propensity operating to reflect how the mind itself must operate in order to detect and respond to reality.

This is the subject of Chapter Sixteen, The Mechanical Mind.

13. RANGES OF SURVIVABILITY: THE THREE STATES OF MATTER

Present concepts put the three states of matter on a level of understanding comparable with knowing that it is the key that starts the car.

Matter is what we can touch.

It is solid, liquid or gas.

However, the matter we are concerned with is physical, chemical and equilibrical.

Physical matter is made up of atoms, with units of elementary particles in the nucleus, and clouds of orbiting elementary particles. It is in balance with the field in which it exists and can survive to the extent that it can exchange elementary particles with the field.

Chemical matter is physical matter that moves elementary particles back and forth internally so as to balance the changing field in which it exists. Forming into chemical matter extends the range of survivability of physical matter.

Equilibrical matter is made up of individual elementary particles locked into stable equilibrium by their opposing properties of affinity propensity and at-rest motion.

The concept of equilibrical matter is induced from how the mind operates and uses the elementary particle with its two properties. It serves as a detector which allows the chemical matter to which it is attached to move through physical reality and therefore extend its range of survivability.

Equilibrical matter can form into the detector itself, the mind, or the microstructures that encode physical reality.

An elementary particle comes into existence to define non-existence.

As additional elementary particles come into existence, they form into units. The units are made up of a precise number of elementary particles. That number is limited by the inability of the affinity propensity, the property that makes each elementary particle attempt to occupy the physical space of the other, to cause one more elementary particle to overcome its at-rest motion of the speed of light, the second property of the elementary particle.

PHYSICAL MATTER, AS OPPOSED TO CHEMICAL MATTER, IS FORMED ATOM BY ATOM IN THE ABSENCE OF A FIELD, AND BREAKS DOWN IN A FIELD OF COMBUSTION IN ORDER TO ACHIEVE A BALANCE WITH THAT FIELD.

The units, do, however, have an excess of affinity propensity as a structure that either causes elementary particles to be attracted to the extent that they will orbit the units, or cause the units to be attracted together into a nucleus, most probably a combination of both, with the orbiting elementary particles establishing orbits shared by the individual units.

Each nucleus, formed in the total absence of a field of electromagnetic emissions, is formed identically, and the resulting structure, still having an excess of affinity propensity, although not enough to attract an additional unit into the structure, still has enough to attract orbiting elementary particles and other nuclei with orbiting elementary particles so that the atoms combine into physical matter.

The atoms form together in all directions, which describes a sphere, so that the resulting physical structure forms into a ball.

When it ignites, it becomes a star such as our sun.

If the ball of matter is the size of the Earth, and ignites at the same time as a ball of matter that makes up the sun, the smaller ball of matter will cool relative to the sun.

If the ball of matter is the size of the sun, or even larger, and it is part of thousands of balls of matter that formed in the center of the sphere of matter formation, then when these spheres begin to ignite, they will begin to emit expanding fields of electromagnetic emissions, which, because at each point contain the attractive mechanism will begin to attract all other suns into one another.

However, because the inductive flows of the expanding emission fields cause the stars to establish rotation around an axis, the stars do not fall toward each other, but rather away from each other.

Each establishes a direction of rotation identical to the

others because of the right hand rule, a result of each cooling as it undergoes combustion after ignition, and each falls away from each of the others in the direction of rotation.

The rotation in turn, establishes an equatorial field of emissions which is stronger than at any other point in the emission field.

GALAXIES ARE FORMED IN MUCH THE SAME WAY THAT THE SOLAR SYSTEM SORTS ITSELF OUT, WITH COMBUSTION ESTABLISHING ROTATION, ROTATION PLANULARITY, AND MATTER MOVING TOWARD THE STRONGEST AREAS OF ATTRACTION IN THE COMBINING EMISSION FIELDS.

As the stars ignite, establish rotation and orbital patterns, and the rotation establishes an equatorial emission field, the stars move as a result of the forces acting upon them, the forces of the attractive mechanism and the inductive flows contained in the expanding fields of emissions, the fields themselves regulated by the rate of the combustion process which in also controlled by the strength of the combining fields of emissions.

As the movement occurs, the stars start to sort themselves out, aligning themselves with each other in the equatorial emission fields established by their common rotation.

The expanding emission fields reach the point in the area of matter formation at which our sun is forming, the sun ignites, and itself becomes a part of the process.

If we ignore the planet that ignites at the same time as the sun, we can focus on what might happen with the sun as it underwent combustion.

It's emission field, its starlight, joins the starlight of the other stars that make up the galaxy. That starlight itself becomes a single emission field, which, at far distance points in the universe looks like a single star.

Eventually the combined emission fields expand to a point at which the affinity propensities holding them together are no longer capable of directing the at-rest motion of the elementary particles so that they contain a coherent structure as the electromagnetic spectrum.

The electromagnetic spectrum breaks down, and the elementary particles become part of the emission fields, the inductive flows of directed flows of elementary particles passing through the region where the break down occurs. If there are no directed flows, the elementary particles restart the process, forming into matter which achieves a physical presence, and combines into a structure until circumstances cause field replacement to begin, the structure ignites, and begins to combust, creating an expanding field of electromagnetic emissions.

The process is cyclical, regulated by the existence of a field of electromagnetic emissions, with matter breaking down in the strongest field possible, and reforming in the absence of a field.

Every piece of matter in existence will eventually go through the cycle because every piece of matter will be drawn by the attractive mechanism embedded in the expanding emission field back toward the source of emission.

If it is not big enough to combust, rotate, and avoid a direct collision, as Halley's comet is still able to do, then it will become a part of the emission field of the sun.

Halley's comet will eventually break up and also become a part of the emission field of the sun.

As Mercury moves around the sun, it heats up at its closest approach, and the increased emissions increase the force of the attractive mechanism embedded in its field. The increase in the attractive force causes it to pop its orbit, in effect achieve an entirely new orbit (a fact used to justify notions that mass, space and time are interchangeable quantities). Each pass causes the elements that make up Mercury to emit and break up so that eventually, when there are no longer elements that can do anything but melt, it will be pulled into the surface of the sun and also become a part of the sun's emissions.

All matter goes through this cycle!

No matter in the universe can escape it!

Matter is formed, it ignites, it combusts.

As it combusts, the elementary particles that make it up depart the matter as their at-rest motion is regained.

They are released in the form of the electromagnetic spectrum.

The electromagnetic spectrum contains stationary affinity propensities which can replace the affinity propensities holding the matter together, causing it to also undergo combustion.

The electromagnetic spectrum also contains the attractive mechanism which draws matter deeper into the electromagnetic spectrum ensuring that it will undergo combustion.

IF THE FORCE OF ATTRACTION WERE THE ONLY FORCE MOVING MATTER, THEN THE ONLY MATTER THAT WOULD EXIST WOULD BE PHYSICAL MATTER, FORMING IN THE ABSENCE OF A FIELD AND MOVING TOWARD COMBUSTING FIELDS TO UNDERGO COMBUSTION IN THOSE FIELDS.

The electromagnetic spectrum resulting from the combustion itself expands to the point that it breaks down. The elementary particles that formed it reform into matter that restarts the cycle.

Physical matter forms from the elementary particles of expanding fields of electromagnetic emissions, moves with respect to expanding fields of electromagnetic emissions which insure that the matter will turn into expanding fields of electromagnetic emission which will become elementary particles from which matter will reform.

This is what happens with physical matter, and if that

were all that happened with physical matter, then we would have a perfect universe, a universe that operated like a perpetual motion machine, filling non-existence with all of the elementary particles that could ever be created, moving those particles in an endless cycle of formation, ignition, combustion, expansion, dissipation, formation, ignition, combustion, expansion, dissipation...

It would be kind of boring, however, and it would have no purpose.

IF A PARTICLE IS GOING TO SOMEHOW COME INTO EXISTENCE TO DEFINE NON-EXISTENCE, THEN IT IS GOING TO HAVE TO BE ABLE TO ARRANGE ITSELF IN A MANNER THAT WOULD ALLOW IT TO COMPREHEND ITSELF, EXISTENCE IN NON-EXISTENCE.

Physical matter, simply by its formation and dissipation, would not be able to comprehend itself.

And there would be little use for the creation of a particle to define non-existence, if the particle itself couldn't comprehend its own existence.

Because physical matter does not comprehend, the elementary particle must have been created with its properties, the affinity propensity and the at-rest motion, in such a way that it can form itself into structures other than physical matter.

Now that we have seen that the two properties synthesized for the elementary particle allow that elementary particle to operate in a manner that provides a consistent picture of physical phenomena, we can take those two properties and see if there is any other structure that they can create.

We have already seen one such structure when we artificially created a flow of elementary particles by creating a difference in potential that allowed the elementary particles to move in a path from a source where there was an excess of elementary particles to a place where there was a deficit of elementary particles, the deficit being evidenced in excess affinity propensity to which the elementary particles moved.

If we completed the analogy of the battery we would see the flow move toward the excess affinity propensity, satisfy that excess by its presence, and move back in the circuit at its at-rest speed to the area of the source, where it would continue to act in a field replacement function, liberating elementary particles to add to the flow.

When we created the battery, we manipulated physical matter to create flows of elementary particles.

When physical matter arranges itself so that it can create flows of elementary particles among its changing potentials as a result of existing in a changing emission field, it has evolved into chemical matter.

What is it in objective physical reality that creates chemical matter, the ability of matter to alter itself with respect to the changing emission field in which it finds itself.

Because if chemical matter cannot create itself in

physical reality, the chemical constructions that make up our bodies, including our brains, could never have sprung from physical reality, and we would not be here attempting to comprehend that physical reality.

Chemical matter is the second state of matter.

What is it that drives the creation of chemical matter?

Physical matter is made up of units of elementary particles in equilibrium. It contains elementary particles moving at their at-rest speed of light, but that at-rest speed is expended in stable orbits around the nuclei, and the units that make up the nuclei.

PHYSICAL MATTER FORMS IN THE ABSENCE OF A FIELD. CHEMICAL MATTER FORMS IN A CHANGING FIELD.

The nuclei have units which are substantial enough to be affected by the attractive mechanism in the expanding emission fields, and therefore be attracted toward the source of emissions.

Chemical matter, on the other hand, is made up of connected units of physical matter with differing potentials which exist in changing fields.

The changing field alters the equilibrium of the physical matter causing changes in potential.

Changes in potential cause flows of elementary particles between and among the physical matter.

Physical matter moves to attempt to reach equilibrium by moving with respect to changing emission fields.

Chemical matter alters its structure as it responds to the changing emission field, establishing flows within itself.

The ability of physical matter to exchange elementary particles with emission fields determines the existence of physical matter.

On the other hand, the flows between the physical matter in chemical matter regulate its existence.

IF PHYSICAL MATTER CAN MOVE WITH RESPECT TO A CHANGING EMISSION FIELD, IT CAN REMAIN AS PHYSICAL MATTER UNTIL IT UNDERGOES COMBUSTION. IF PHYSICAL MATTER CAN NOT MOVE WITH RESPECT TO A CHANGING EMISSION FIELD, THE CHANGING EMISSION FIELD WILL ALTER THE PHYSICAL MATTER, TURNING IT INTO CHEMICAL MATTER.

Thus, in physical matter, the equilibrium is between the matter and the field, while in chemical matter the equilibrium exists in the movement of the elementary particles among the changing potentials of the physical matter that makes up the chemical matter, and then with the field.

In both cases, equilibrium is the goal.

In physical matter, equilibrium is obtained only by fixing the object with respect to the field.

In chemical matter, equilibrium can never be achieved because the object can never be fixed with respect to the field.

Chemical matter is in a constant state of disequilibrium because the field in which it exists is constantly changing.

The only way that it can survive in a changing field is to be in a position to alter its equilibrium so that it will constantly be in equilibrium with the changing field in which it exists.

If you move physical matter in the field, the physical matter will merely attract elementary particles out of the field, or emit them into the field.

It has no other choice.

If it emits too many, it will be destroyed. It can not adjust to the field in which it exists.

If you move chemical matter in a field, the chemical matter will adjust to the field by allowing flows to internally move back and forth so as to balance the changing external field.

The absolute distinction, then, is that the ability of physical matter to form into chemical matter extends its range of survivability.

Physical matter, as mere atoms, are directly dependent on their position in the emission field for their survivability.

Because of the attractive mechanism embedded in that field, physical matter will inevitably be drawn into a combustion field so that it will itself ignite and undergo combustion.

Chemical matter is physical matter that has been arranged into a structure that allows it to move flows of elementary particles among its atoms of physical matter so that when those atoms are moved with respect to the external emission field, they can readjust their affinity propensities internally so that there will be a range of existence over which the external emission field cannot effect their existence.

CHEMICAL MATTER REDISTRIBUTES ITS INTERNAL AFFINITY PROPENSITIES TO ACCOMMODATE CHANGES IN THE EXTERNAL EMISSION FIELD. THE CHEMICAL MATTER THAT MAKES UP OUR BODIES MAINTAINS ITSELF AT A STEADY TEMPERATURE REGARDLESS OF THE FLUCTUATIONS OF HOT AND COLD IN THE EXTERNAL FIELD. THE CONVERSION OF PHYSICAL MATTER INTO CHEMICAL MATTER EXTENDS ITS RANGE OF SURVIVABILITY.

In physical matter the affinity propensities and at-rest motion are fixed.

In chemical matter, the affinity propensities and at-rest motion move over a range so that they can adjust to the fixed nature of the physical matter and the field.

Is there a third arrangement that could form out of elementary particles with at-rest motion and affinity propensity which would extend the range of survivability of the elementary particles?

Chemical matter, in spite of its ability to adapt to the changing emission fields in which it finds itself, is still subject to the attractive mechanism embedded in the emission field.

Sooner or later, chemical matter is going to end up undergoing combustion because the moving flows of elementary particles merely adjust the potentials in the physical matter to increase its range of survival.

Chemical matter is still physical matter with units that are subject to the attractive mechanism.

The flows will become part of the emission field itself, if the emission field is strong enough, and the atoms without the flows are just atoms of physical matter.

Is there another structure of matter that the elementary particles with their at-rest motion and their affinity propensity will naturally form?

Would such a structure be transparent to the attractive mechanism and thus be able to extend its range of survivability even further?

Far enough, in fact, for it to develop comprehension?

The attractive mechanism results from the interaction of the expanding flows of elementary particles. The flows break up, overexpand, intermingle with one another, and as in interference, the inductive flows created by their presence recombine them, capturing the unit of the atoms in the force of recombination, and causing them to move back toward the source of the flows.

This mechanism is made up of elementary particles and acts directly on the units of the atoms that make up matter because of the units physical size.

Any elementary particles that are in the area of the attractive mechanism will be there because they make up the unit, which is stationary with respect to the matter it is a part of, or they are a part of the flow, in which case they are structured into the flow, and depart as soon as they arrive, or they are a part of the inductive flows recombining the intermingling flows.

BOTH PHYSICAL AND CHEMICAL MATTER ARE SUBJECT TO THE COMBUSTION CYCLE, BEING CONSTRUCTED SO THAT THE ATTRACTIVE MECHANISM CAN AFFECT THE UNITS OF THE NUCLEI OF THE INDIVIDUAL ATOMS, DRAWING THE MATTER INTO THE COMBUSTION PROCESS.

Any other elementary particles would simply not be in the area of the attractive mechanism because the attractive mechanism itself could not physically influence their motion and the affinity propensities in the area could not hold them in place.

But what if an elementary particle were to be forced to remain stationary with respect to the attractive mechanism?

Would the attractive mechanism have any effect on it?

No, because the attractive mechanism itself is made up of elementary particles. A single elementary particle will be transparent to the attractive mechanism which is itself made up of individual elementary particles.

If the elementary particles could somehow form into a structure in which the individual elementary particles were stationary with respect to the structure, but still individual so that they would be transparent to the attractive mechanism that causes units of elementary particles to move back toward the emission field, then the range of survivability of the structure would be vastly increased.

IN ORDER TO DEVELOPE THE PERMANENCE NEEDED TO COMPREHEND ITSELF, CHEMICAL MATTER WOULD HAVE TO BE CAPABLE OF DEVELOPING A STRUCTURE WHICH IS NOT SUBJECT TO THE COMBUSTION CYCLE.

The structure would not be attracted to, and become a part of combusting matter.

It would be removed from the cycle of formation, ignition, combustion, emission, and dissipation.

It could, in fact, stand outside of that cycle so that it could comprehend both the cycle and itself.

What would be the nature of such a structure?

The property of at-rest motion of the elementary particle is always in opposition to the property of affinity propensity.

The elementary particle has a need to move at its at-rest motion that is perfectly balanced by its need to occupy the space of all other elementary particles, its affinity propensity.

In physical matter, the affinity propensity has overcome the at-rest motion.

In the electromagnetic spectrum the at-rest motion has overcome the affinity propensity.

There only remains the situation in which the at-rest motion exactly balances the affinity propensity.

Such a situation could exist if the elementary particles were put into a structure so that they were evenly spaced from one another.

Because the affinity propensity opposes the at-rest motion, to the extent that the affinity propensity overcomes the at-rest motion, the potential for the at-rest motion to regain its at-rest speed increases.

ATOMS TODAY ARE CONCEIVED TO BE HELD TOGETHER IN STABLE WEBS OF FORCE. IF ELEMENTARY PARTICLES WITH OPPOSING PROPERTIES ARE SUBSTITUTED FOR ATOMS, WE HAVE THE CLASSIC STRUCTURE MAINTAINED BY OPPOSING FORCES: A PARTICLE SITUATED IN THE WEB OF EVENLY SPACED ELEMENTARY PARTICLES COULD NOT MOVE TOWARD ANY SINGLE AFFINITY PROPENSITY WITHOUT ITS AT-REST MOTION PULLING IT BACK INTO ITS ORIGINAL POSITION IN THE WEB OF FORCE.

If elementary particles were somehow made to be evenly spaced in a structure, then the affinity propensities of the structure would be balanced, each particle being equally distant from all other particles and therefore incapable of moving towards any.

However, the affinity propensities would not be entirely satisfied, no one particle occupying the space of another.

The at-rest motion of the elementary particles locked together in such a structure, though, would be totally overcome, and therefore their potential for regaining their at-rest motion is as high as it is when they are clumped together into a unit at the nucleus of an atom.

Such a structure would lock the elementary particles in a solid embrace of their opposing properties, with any movement of an elementary particle toward another elementary particle releasing a portion of the at-rest motion which would pull the elementary particle right back into its normal position within the structure.

Such a structure would be the only additional structure permitted by the two properties of the elementary particle, its at-rest motion and its affinity propensity.

Such a structure would be virtually indestructible with respect to the formation, ignition, combustion, emission and dissipation cycle of physical matter. It would not be affected by the attractive mechanism which draws physical matter into the combustion process for reduction to its basic elementary particles.

Once the elementary particles had been locked into position, no external affinity propensities could affect the overall structure because any movement of one elementary particle toward the affinity propensity of another would cause its at-rest motion to move it right back to its position in the structure.

Such a structure would be in equilibrium with itself.

Any disbalance that might be caused by the affinity propensities of flows passing through it would be countered by its internal dynamics. The movement of particles toward the presence of the flow would be balanced by the release of at-rest motion moving the particles back into their position in the equilibrical structure.

What could produce this equilibrical state of matter?

WE KNOW THAT CHEMICAL MATTER IS PRODUCED BY THE INTERACTION OF PHYSICAL MATTER IN CHANGING FIELDS. WE ARE CHEMICAL MATTER IN A WORLD MADE UP OF PHYSICAL MATTER. EQUILIBRICAL MATTER IS THE NORMAL DEVELOPMENT OF CHEMICAL MATTER THAT NEEDS TO DEVELOPE THE ABILITY TO PERCEIVE THE ENVIRONMENT IN WHICH IT HAS TO MOVE TO SURVIVE.

As the flows in chemical matter modulate the affinity propensities of the physical matter that makes up the chemical matter, they create affinity propensities that could immobilize the at-rest motion of a single elementary particle.

If physical matter can evolve chemical matter, chemical matter can evolve equilibrical matter.

It is easier to see how the affinity propensities in chemical reactions can contain the at-rest motion of a single elementary particle so as to snap it into a structure that will be equilibrical than it is to see how physical matter can evolve chemical matter.

To see how physical matter evolves chemical matter, we don't have to figure out something as difficult as why the sun rises.

We only have to know that it does.

14. THE ABILITY NOT TO ACT: COMPREHENSION

All matter forms into atoms with a nucleus containing the highest number of units that can form together as a result of the affinity propensity immobilizing the at-rest motion of the individual elementary particles.

Because the Earth was formed out of atoms with the highest number of units in their nuclei, when it ignited, the atoms that were igniting on the surface of the sphere were atoms with the highest atomic numbers that can exist.

But the Earth didn't ignite in isolation. It ignited next to a dominate emitter, the sun. As such, its rate of combustion was greater than it would have been had it ignited in isolation.

CHEMICAL MATTER EXTENDS THE RANGE OF SURVIVABILITY OF PHYSICAL MATTER. EQUILIBRICAL MATTER EXTENDS THE RANGE OF SURVIVABILITY OF CHEMICAL MATTER.

What effect did this have on the surface of the Earth?

At the beginning, the Earth acted like a comet with respect to the sun. It was pulled into its field, burned at a faster rate than it would ordinarily burn, established spin, passed by the sun, and moved with its momentum out into space, into the lesser emission field of the sun.

When it made its closest approach to the sun, its surface would be reduced to a boiling inferno of what detectors would measure to be hydrogen atoms, the atoms stripped down to their individual units.

As the Earth moved away from the sun, its surface would be coated with these individual units, their excess affinity propensities increasing as the sun's field decreased.

This resulted in the units on the surface combining together to form elements that could exist in equilibrium with the decreasing emission field.

When the attractive mechanism embedded in the sun's expanding emission field finally overcame the Earth's momentum, the Earth once again began its plunge toward the surface of the sun.

The new elements making up its surface once again began to boil in the sun's heat, breaking down into the units of matter that made them up.

The Earth would pass the sun once again, and move back into space.

The process was complicated by the fact that there were other spheres of matter undergoing the same process as they all attempted to establish orbits around the dominate emitter that would allow them all to move at rest at equilibrium with each other.

The net result was that the surface of the emitter that would become the Earth eventually crusted over with the elements that had formed from the repeated cycles of heating and cooling.

These elements were the elements up to the atomic number of the atoms that could exist in the emission field in which the Earth obtained its equilibrium.

The emission field is the combination of the field of the dominate emitter, the sun, and the emissions of the Earth as it undergoes the cooling process.

This process establishes a continuum for the boundary area between the space in which the matter exists and the combusting atoms, the surface of the Earth.

The surface is cooling slowly, with the sphere of combustion caused by the breakdown of the atoms with the highest atomic number, the atoms that make up the majority of the sphere, providing a second surface, receding away from the surface toward the center of the planet as the planet cools.

AS THE EARTH COOLED, ELEMENTS FORMED FROM THE BREAKDOWN OF THE MORE COMPLEX ATOMS, RESULTING IN THE VARIETY OF PHYSICAL MATTER FOUND IN THE EARTH'S CRUST.

The makeup of the Earth is not determined by the Earth's movement in space, which is an orbit in equilibrium with all of the other emitting matter making up the solar system. The Earth is a sphere with atoms in the center with the highest atomic number possible, surrounded by a sphere of intense combustion which represents these atoms breaking down into atoms with lesser atomic numbers, surrounded by the elements with the lesser atomic numbers cooling toward the surface, a surface constantly replenished with the atoms being broken down.

The physical matter falling out from the breakdown of the more complex atoms in the combustion process would then be affected by rotation.

Rotation is the other force that served to mix up the surface of the Earth as it was crystallizing into elements of matter.

If the planet were stationary, rather than rotating, then the units of matter recombined on its surface might contain the primordial elements only, carbon, chlorine, calcium, hydrogen, iron, magnesium, nitrogen, oxygen, phosphorus, potassium, sulfur and sodium, the elements with lower atomic numbers, fewer units in the nuclei, the elements suitable to life.

THE PHYSICAL MATTER THAT MAKES UP LIFE IS IDENTIFIABLE IN THE ENVIRONMENT. WHERE DID THE ENERGY COME FROM THAT ALLOWED THE PHYSICAL MATTER TO FORGE THE CHAINS OF LIFE?

These elements might be suitable to life, but they would produce no life, because life, animate matter, requires the flows of elementary particles that are the hallmark of chemical matter, and a soup of the primordial twelve will not allow established flows to be created.

Fortunately, the elements with higher atomic numbers were congealing beneath the primordial soup, and the rotation of the planet settled down as the solar system moved into equilibrium.

The stress on the forming crust of the planet speeding

up and slowing down with the changing emission fields in which it found itself caused these elements with higher atomic numbers to the surface to form the surface of a solidifying planet.

As the planet settled into orbit, the crust solidified, and the orbital relationship with the sun established a settled rotation, these elements created a surface which alternated from being in the heat of the sun to being in the cool of the shade.

The planetary rotation established night and day.

Thus, the surface of the planet came to be composed of vast fields of randomly distributed elements which moved on a uniform basis from hot to cold and back to hot again, only to move once again into cold.

THE ENERGY FOR LIFE COMES FROM THE ROTATION OF THE PLANET IN FRONT OF THE SUN WHICH ESTABLISHES FLOWS OF ELEMENTARY PARTICLES AMONG THE CHANGING POTENTIALS IN THE PHYSICAL MATTER OF THE EARTH AS IT CHANGES TEMPERATURE BETWEEN NIGHT AND DAY.

We know that elements are matter, and that matter exists with respect to the emission field in which it finds itself.

Matter also has its own unique structure, being in equilibrium with respect to a field of emission that varies from one element to the next.

We saw when we constructed our battery that two

elements could be in equilibrium with respect to a field, and not be in equilibrium with respect to each other.

We labeled this difference as the difference in potential which produces an electromotive force between the two elements. If a channel is provided between the two elements, an electrical flow will result.

Another thing that we know about this difference in potential is that it alters with each element with the change in the temperature of the element, or in our terms, with the change in the field in which the element finds itself.

THE IMPORTANCE IS NOT IN KNOWING HOW THE SUN COMES UP, THE RESULT OF INDUCTION INDUCED ROTATION, BUT RATHER THAT ITS RISE REFLECTS THE ROTATION WHICH CREATES THE CONSTANTLY CHANGING FIELDS THAT PRODUCE THE CHEMICAL MATTER OF LIFE. SUNRISES ARE LIFE.

The excess affinity propensities of elements differ with respect to each other in a given field, and that difference changes with a change in the field.

Therefore on the forming Earth, if we have two elements lying side by side, and the Earth is not moving, but is rather stationary with respect to a field, the positioning of the elements will eventually result in an exchange of elementary particles with each other and the environment until a point is reached at which both are in equilibrium with respect to one another.

However, the Earth is not stationary with respect to the field in which it finds itself, it is in fact in constant motion, rotating in front of burning sun, so that the elements that are lying side by side randomly over its surface are constantly subjected to extremes of temperature, first heating up in the burning sun, next freezing in the midnight cold.

Equilibrium is impossible!

There can be only a continuing attempt to achieve equilibrium, the continuing attempt we have posited that defines chemical matter.

As the Earth rotates, then, elementary particles are going to constantly be moving back and forth between the elements on its surface in an attempt to balance their excess affinity propensities.

Such a movement should grind the matter into dust.

However, it doesn't.

Why?

Because of the elements that make up the primordial soup.

The models that are devised to construct the monomers, the molecules that are the building blocks of living systems and are formed out of this primordial soup, get hung up on the fact that monomers require energy to link into polymers. The source of energy for life is found in polymers, so how do you get polymers without polymers?

The energy source for polymerization is not found in the primordial soup, but rather in the electrical flows between the physical matter on the surface of the Earth.

These electrical flows are established by the changing potentials in the physical matter, the existence of the affinity propensities of the physical matter with respect to the emission field.

The changing potentials result from the Earth's rotation in front of the sun.

The rotation results in a changing emission field.

The changing emission field changes the affinity propensities of the physical matter.

The changing affinity propensities result in flows between the elements with the changing potentials.

The flows between the different elements will take the course of least resistance, on the surface of the planet.

As the flows move, they create secondary flows, or inductive flows.

As the Earth's rotation stabilizes, the flows stabilize.

The primordial elements are contained in the gaseous soup above the rotating planet. They are subject to the same changes in emission field as the elements on the surface of the Earth. However, with no fixed points between the changing potentials, they do not stabilize.

They therefore attempt to obtain stability by organizing themselves around the electrical flows moving across the surface of the Earth.

The atoms thus form physical structures on the flows between the changing potentials in the elements on the surface of the Earth. These physical structures grow into visible matter that is distinguishable from physical matter in that it is animated by the flows of elementary particles.

The resulting chemical matter that we measure as life provides a path through which the flows between the changing potentials in the physical matter can be modulated.

LIFE IS THE SPONTANEOUS ORGANIZATION OF ATOMS AROUND THE STABLE FLOWS MOVING OVER THE SURFACE OF THE EARTH AS A RESULT OF ITS ROTATION IN FRONT OF THE SUN.

Life forms spontaneously around the affinity propensities of flows of elementary particles that move between changing potentials on the surface of the planet as a result of the Earth's rotation in front of the sun. By providing a way for the physical matter to obtain and release elementary particles on a local basis, it extends the range of survivability of the physical matter.

Its early evidence is in the crystalline like lichen that forms spontaneously on barren rocks.

But lichens are not really animated, or at least not unless they occupy the moving flows around which they form, and from which they obtain their stability.

Remove the flows from lichen and they become nothing more than physical matter.

The passive dampening of destructive flows between the changing potentials of physical matter does not provide comprehension.

However, if lichen were to develop into something else, matter that satisfied the changing potentials in the environment by moving the elementary particles back and forth within itself, rather than exchanging them with the environment, then the animate matter would no longer be dependent on its location for its animation, but could be animate independent of location.

THE DIFFERENCE BETWEEN AMBULATORY MATTER AND NON-AMBULATORY MATTER IS THE ABILITY OF AMBULATORY CHEMICAL MATTER TO RESPOND TO ENVIRONMENTAL CHANGES BY USING FLOWS OF ELEMENTARY PARTICLES TO ALTER INTERNAL AFFINITY PROPENSITIES IN ORDER TO REMAIN IN EQUILIBRIUM WITH THE ENVIRONMENTAL CHANGES.

Animate matter could become ambulatory.

When we see the tall trees reaching up into the heavens, we can visualize the structures serving as a path for the exchange of elementary particles between the Earth and the atmosphere as the Earth rotates from night to day and back to night again.

We would hardly expect the trees to get up and walk away.

Flowers might open up in sunlight to dissipate elementary particles into the particle starved inductive flows, only to close up at night as the surface cools off, draining every elementary particle it can from the surface in response to the increase in excess affinity propensities as the sun's emission field fades.

But if we extirpate it, so much for the animation.

Chemical matter is animate matter that is one step removed from physical matter. If the animate matter is not ambulatory, it cannot develop comprehension.

Comprehension is the result of the need for chemical matter to detect the environment so that it can move within that environment in a manner that will extend its range of survivability.

However, to move within the environment, chemical matter has to develope a method of internally balancing the external changes in the environment.

As the Earth's crust cooled into rocks, and life formed on the rocks, the elements in the primordial soup began to reach stability with one another, with the abundant hydrogen combining with the abundant oxygen when the emission field made it possible for the two to be stable together.

As more than half of the Earth's surface became covered with water, the need for matter that could exist in changing potentials, independent of the flows established by the potentials became a necessity.

The resulting structure, the cell, can be analogized to vegetation that develops the ability to satisfy the difference in changing potentials by moving elementary particles back and forth within itself, rather than passing them back and forth between itself and the physical matter.

Instead of passing the elementary particles into the physical matter that was undergoing the change in potential, the structure simply moves elementary particles back and forth within itself in response to the changing potentials, satisfying the changing potentials in the physical matter in the process.

Such a structure could not develop on the elements that make up the surface of the Earth because lichen is dependent on the flows for its existence.

Lichen cannot leave the flows to develop a system that works independently of the flows, because without the flows lichen become physical matter.

However, come the rains, with the lichen being continuously washed into the forming ponds, lakes, rivers and oceans, the lichen had to adapt into a cell that internally balanced the changing potentials in its environment.

The changing environment of the water could not itself develop the cell because the environment was not stable enough to maintain stable flows. It required structures that developed in response to stable flows on land to adapt to the changing flows in the new environment.

The structure could start out like the chemical matter that made up lichen, being a passive channel for changes in potential. However, in water, the chemical matter would form into a structure that moved the elementary particles back and forth within itself rather than back and forth between two elements within the environment.

We can see this movement as the walls of the cell move this way and that way in response to the changing fields of its environment.

As it moves within that environment, it dissipates elementary particles into the environment. It therefore has to obtain elementary particles out of the environment. To be effective, it has to be a net receiver of elementary particles to maintain its stability.

THE NEED FOR AN EQUILIBRICAL STRUCTURE DEVELOPED WHEN ANIMATE MATTER DEVELOPED THE ABILITY TO MOVE INDEPENDENT OF ITS ENVIRONMENT.

As a net receiver of elementary particles, the cell periodically has to divide in order to survive.

Such a cell is dependent on its random movement within the environment.

A cell that moved randomly would not require a structure to detect its environment.

A cell that could move independent of its environment

would extend its range of survivability.

The formation of life and comprehension is a natural progression from the interaction of matter in physical reality.

The combustion process breaks down matter into simpler elements.

Rotation provides the flows around which the simpler elements develope into chemical matter.

Ambulatory matter can not develope in a primordial soup because there are no defined points between which the flows can be established.

However, once chemical life develops on the rocks, it can adapt to ambulatory conditions when the soup liquifies.

Once in the oceans, survival favors cells that can move independent of the environment rather than just react to it.

To move independent within the environment, the cell has to develope a method of detecting the environment.

The only thing in the environment to detect is expanding spheres of electromagnetic radiation: Heat and light.

One way for an organism to sense the environment is to utilize the changing potentials of physical matter to determine the strength of the field. As the matter's existence is dependent on the strength of the field, the organism develops a detector to count the number of elementary particles that specific matter releases or captures as a result of the changing fields in which it exists.

Because the emission field expands inversely with the square of its distance from the source, an organism can use this direct field detection either to move toward the emission field, or move away from the emission field.

We can accomplish the same thing by shutting our eyes and moving our fingers toward a match. We know how close we are to the match by the changing affinity propensities in the matter that makes up our nerve endings.

> **CHEMICAL MATTER CAN SENSE CHANGES IN THE ENVIRONMENT BY FIELD REPLACEMENT, LOCATING ITSELF WITH RESPECT TO THE EMISSION FIELD BY DIRECTLY DETECTING THE STRENGTH OF THE EMISSIONS.**

Whatever area of the brain that controls our reaction, the reaction is certain. It will cause our hand to move away from the flame if the field is too strong for the nerve endings.

Once mobility is obtained, once the organism becomes ambulatory with respect to its environment, it needs something more than an emission field for it to operate within the environment.

The strength of the emission field doesn't tell the organism enough about the environment.

If it is to survive, it needs more accurate information. If

we eliminate taste and smell as being inadequate to provide an organism with the ability to move freely within the environment, we are left with sound and sight.

Sound is generated in a medium, and produces waves with respect to the medium that diminish with their distance from the source.

Sound, therefore, can tell you a lot about the environment.

We have all seen a squirrel cross a street, stopping at every noise to adjust its course in order to proceed safely along to its destination.

A loud sound in the environment will bring an animal to a temporary halt. It doesn't want to jump one way or the other until it can identify the nature of the sound, because if it does, it might jump right into the way of the harm the sound represents.

Whatever the area of the brain that is controlling the animal's reaction, the reaction is certain.

The animal comes to a dead halt.

What is going on during this interval?

If the animal had come into contact with something hot, it would have jumped away instantly. The reaction is wired into the system.

Here, chemical matter is indirectly detecting the environment. The sound is picked up by the ear, converted to an electrical flow, and the electrical flow is carried to the detector.

Once the flow is established, the detector works on the same basis that the direct field detector works. Field replacement occurs in specific physical matter, and a detector measures the rate at which the elementary particles are released.

The louder the noise is, the greater the flow, the greater the field replacement, and the greater the number of elementary particles released and detected.

What is going on in the brain in the interval between the sensing of the noise and the ability to act in response to the noise?

And where is it going on?

The animal has to connect the sound with the environment in order to move or not move in a manner that will extend its range of survivability within that environment.

To connect the sound with something in the environment, it has to have the ability to reconstruct that environment in its brain.

It has to then identify what it is in that environment that is producing the sound.

And then and only then, after it has comprehended what it is in the environment that is causing the loud noise, can it form a picture of itself acting or not acting in that environment in a manner that will extend its range of survivability.

A lot was going on in that interval.

It has to reconstruct a picture of the external environment in which it exists.

It has to find out what it is about that environment that is providing a warning signal, that is making the environment something other than what it should be.

It has to put that warning signal into the context of its own existence in the environment.

It then has to form a picture of itself acting or not acting in a manner that will allow it to continue to exist within that environment.

Only then can it act, or not act in a manner that will allow it to continue to exist within that environment.

The noise in the environment can stop the elementary particles that make up the animal from moving in the environment.

The noise stops the animal so that it can survive.

The process requires that the elementary particles comprehend the environment.

How does this process of comprehension work?

What is the nature of a structure of elementary particles that will allow chemical matter to detect the environment so that it can move safely within that environment?

15. THE ROOT OF COMPREHENSION: COMPARISON

The mind operates on the basis of comparison. We compare the picture of external reality in which we find ourselves with a normal state that reflects that reality, if only to the extent of accepting it passively. If something in the reality doesn't accord with the state we are using as a benchmark, then the failure to compare catches our attention.

IF WE CAN EXPLAIN HOW WE PERCEIVE REALITY WITH THE SAME ELEMENTARY PARTICLE, WITH ITS AT-REST MOTION AND AFFINITY PROPENSITY THAT WE USE TO EXPLAIN PHYSICAL REALITY, THEN WE WILL HAVE SYNTHESIZED AN ELEMENTARY PARTICLE THAT CAN COMPREHEND ITSELF.

We halt and examine the failure of comparison until we can put the failure to rest, and begin to once again move normally in reality.

We have synthesized an elementary particle with two properties, at-rest motion and affinity propensity, and have been able to use that particle to explain in a consistent manner the physical phenomena in which we find ourselves, and of which we are aware.

If, using that same elementary particle, with the same two properties, we can provide an explanation of how we are aware of the physical phenomena, then we should have some degree of certainty that the elementary particle with its two properties is at the root of both the physical reality in which we find ourselves, and ourselves, the "I" that makes us comprehend that physical reality.

It's always best to seek understanding in the simplest model possible.

If we want to model how chemical matter might evolve a reflexive mechanism, a mechanism that would allow it to communicate with its parts, we don't have to look further than the principle of field replacement.

In the battery we created, we had a place where there was not an excess of potential with respect to the place where there was an excess of potential.

The place where there was not an excess of potential provided the elementary particles, which, under specific chemical conditions, were able to move at their at-rest motion from one place to another along a path provided for that purpose.

When we completed the battery with a circuit, the elementary particles continued on until they came back in the circuit to their source, at which point, because they were the directed field of electricity, they continued to replace what had become an excess affinity propensity.

Just as matter exists relative to the expanding emission field of a combusting body, so does it exist with respect to the directed flow of elementary particles in which it finds itself.

Atoms may be in equilibrium with a large number of elementary particles in orbit around the nucleus.

When a directed field, made up of flows of elementary particles, flows with presence at any point in the flow, is passed by the atom, the orbiting elementary particles of the atom are released to the extent that the affinity propensity of the directed flow replaces the affinity propensity of the orbiting elementary particles.

If the directed field is removed, the atom regains its orbiting elementary particles out of the environment.

We accomplish this effect when we set up a circuit with a battery and turn it on and off, on and off, on and off. When the battery goes on, field replacement releases elementary particles from their orbits around the atoms' nuclei. When it goes off, the atoms regain the elementary particles into orbit.

A DIRECTED FLOW OF ELEMENTARY PARTICLES HAS PRESENCE, AND THEREFORE EXERCISES THE SAME FIELD REPLACEMENT AS AN EXPANDING FIELD OF ELECTROMAGNETIC EMISSIONS.

Chemical matter forms as a result of the changing fields in which it exists. It senses the environment by creating a chemical battery which can emit and capture elementary particles to reflect the strength of the surrounding fields.

All it needs is a detector to quantify these flows.

Equilibrical matter, the third state of matter, forms as a result of the need of chemical matter to move in a changing environment. Equilibrical matter serves as the detector that allows chemical matter to form a picture of external reality.

All that is needed is a switch and a load, the switch being the mechanism that creates the directed flows that will replace the orbiting elementary particles in the atoms, and the load being the effect, the light bulb that emits the elementary particles that will be detected by the equilibrical matter.

We have seen the simplest switch actuated by a loud noise.

THE PROCESS OF SENSING THE ENVIRONMENT IS A PROCESS OF ESTABLISHING A FLOW OF ELEMENTARY PARTICLES IN THE ORGANISM THAT REQUIRES THE KNOWLEDGE OF THE ENVIRONMENT.

The noise enters the receptor, say an ear, causes a membrane, the ear drum, to vibrate. The vibration produces an electrical flow.

The ear drum could be any of the sensory organs.

Once the flow is induced, it is channelled to the load that controls the response to the noise.

The load is physical matter made up of atoms with orbiting elementary particles that are readily available for release as a result of field replacement.

The directed flows generated by the sound, and proportional to the presence of the sound as evidenced in the strength of the sound waves, replace a proportional amount of orbiting elementary particles.

The organism could have a series of detectors to determine the relative importance of the sound to the organism's existence in reality. If the noise is slight, the elementary particles would be few and might be absorbed by the organism without notice.

As the noise increases, the flows it creates increase, increasing the field replacement proportionally, and incrementally affecting a broader range of the subsystems.

The field replacement step is required to amplify the flow and provide the variety of responses necessary for survival.

THE FLOWS OF ELEMENTARY PARTICLES ARE DIRECTED TO PHYSICAL MATTER WHICH EMITS ELEMENTARY PARTICLES AS A RESULT OF FIELD REPLACEMENT.

The range of response could be a gradual change of course as a result of a rising noise to a complete halt as a result of a sudden overpowering noise.

A squirrel running along the street can be brought to a dead stop by the pop of a tire on a stone. A human walking across the floor of a shopping mall can be brought to a dead halt by the sight of a bare chest.

The detector is detecting a normal level of sound in the environment before the sound level changes.

When a normal level of electrical flow is passing through the specific mixture of physical matter, the organism can move through reality without that reality intruding on the physical systems moving the organism through that reality.

With a normal flow, the physical matter is in equilibrium, not releasing elementary particles.

THE FIELD REPLACEMENT FUNCTION BOTH AMPLIFIES THE SIGNAL AND, BY ACTING ON A VARIETY OF PHYSICAL STRUCTURES, REDIRECTS IT TO APPROPRIATE SUBSYSTEMS. ONE LEVEL OF SOUND, A SOFT VOICE, COULD CREATE FIELD REPLACEMENT IN ONE SPECIFIC MIXTURE OF PHYSICAL MATTER AT A LEVEL THAT TRIGGERS THE MUSCLES TO RELAX WHILE ANOTHER LEVEL OF SOUND, A HARSH VOICE, COULD CREATE A LEVEL OF FIELD REPLACEMENT IN ANOTHER SPECIFIC MIXTURE OF PHYSICAL MATTER WHICH RESULTS IN TENSENESS.

This is an equilibrical state.

However, when something in that reality does not compare with the equilibrical state of the specific mixture of physical matter, a loud noise where there should be no noise, a bare torso where there should be no bare torsos, the detector is bombarded with an altered flow.

The failure to compare creates a directed flow which effects the internal subsystems in specific ways.

It is easy to see why a loud noise might bring a squirrel to a halt.

It is not so easy to see how a bare breast might bring a human to a halt.

This bringing of the organism to a halt is actually what is described as getting the organism's attention.

The noise simply changes the state of the equilibrical structure doing the detecting.

The comparison is inherent in the change.

How does the mind compare?

If we use the neural net concept that is popular today, we have the mind comparing itself with its own changed state. In the neural net model, external stimuli light up specific neurons in a pattern. If we see the milky way, our neurons light up the milky way. If we see a brick wall our neurons light up a brick wall. If we see the milky way again, the milky way neurons light up.

ATTENTION, THE FAILURE TO COMPARE, IS THE RESULT OF THE DETECTION OF EXTERNAL REALITY DISBALANCING INTERNAL EQUILIBRIUM. IT PROVIDES THE ORGANISM WITH THE ABILITY NOT TO ACT, WHICH ALLOWS IT TO RECOGNIZE THE ENVIRONMENT AND MOVE SAFELY WITHIN IT.

One of the problems with the neural net notion is that there is nothing to make the comparison. What knows that the random connections established in response to the milky way represents the milky way.

We know about the milky way. We don't just respond to existence like a squirrel responds to a noise.

We need a place in which the comparison can take

place.

We found that there is an equilibrical structure that the elementary particles can exist in if they are placed together in a manner in which their at-rest motion balances their affinity propensities.

If the affinity propensities of elementary particles are used to immobilize their at-rest motion in a structure in which the affinity propensities are evenly distributed, then if something affects the affinity propensity of any one elementary particle, its movement toward that affinity propensity would be counteracted by its at-rest motion working against the affinity propensity, causing the elementary particle to remain in position.

THE BASIS OF COMPREHENSION, COMPARISON, REQUIRES A PLACE FOR THE COMPARISON TO TAKE PLACE.

If we visualize a cube of elementary particles, ten on a side, we would have a structure with a thousand elementary particles, all evenly spaced. This will give us an idea of the equilibrical structure we will be dealing with.

If we wondered what held the elementary particles into their places, we could visualize each particle as being held by the affinity propensities of each of the other particles. All of the particles would be equally distant from the others, so that each would be attempting to occupy the space of all of the elementary particles around it equally, and therefore could not move in any one direction.

Such a structure, like the unit in the nucleus of an atom, would render the at-rest motion of the elementary particles motionless.

However, unlike the immobilization of the at-rest motion in the unit of the nucleus of an atom, the at-rest motion in the elementary particles that make up the equilibrical structure are in balance with the affinity propensities of the particles.

EQUILIBRICAL STRUCTURES ARE THE STRUCTURES FORMED OUT OF ELEMENTARY PARTICLES WHOSE AT-REST MOTION HAS PERFECTLY BALANCED THEIR AFFINITY PROPENSITIES. THEIR FORCES ARE IMMOBILIZED IN A STABLE WEB OF EQUILIBRIUM.

If a particle is disbalanced by moving closer to one of the elementary particles than another, then, when the cause of the disbalance is removed, the particle would immediately move back into its position with the affinity propensities balanced.

Another difference between the equilibrical structure and the unit in the nucleus of an atom is that the size of the unit of an atom is controlled by the affinity propensity using itself up immobilizing the at-rest motion.

In the equilibrical structure, the at-rest motion balances the balanced affinity propensities, and therefore there is no limit to the size of a structure that could develop, so long as the chemical process that created it by immobilizing individual elementary particles continues to provide it with additional particles.

In addition, like the unit, equilibrical structures contain uncountable numbers of elementary particles, the elementary particles being no larger than needed to define non-existence, with the number capable of fitting on the head of a pin being uncountable.

However, because at any point in the equilibrical structure its existence is a point only large enough to define non-existence, it is not itself directly measurable, measuring devices requiring at least a stream of elementary particles one particle in presence to detect matter.

EQUILIBRICAL STRUCTURES ARE CAPABLE OF BEING DISBALANCED, AND ARE THEREFORE CAPABLE OF DETECTING WHEN INPUT DOESN'T AGREE WITH THE BALANCE. SUCH A STRUCTURE THEREFORE HAS THE ABILITY TO COMPARE.

And finally, the attachment of such an equilibrical structure is accomplished in the chemical matter of the brain simply by creating stable areas with excess affinity propensity, the stable areas defining and in turn stabilizing the borders of the equilibrical structure.

If one of the elementary particles that is at rest within an equilibrical structure is moved, it will return to its position of balance as soon as the force that is causing it to move is removed.

The elementary particles that make up the equilibrical structure are in balance, and if something disbalances them, they will regain their balance as soon as whatever is disbalancing them ceases to disbalance them.

What could disbalance the elementary particles in an equilibrical structure?

A directed flow of elementary particles, moving through the equilibrical structure, would create a presence within the structure because at every point in the flow there would be an elementary particle as one left the point and the next one occupied the point.

BECAUSE THE EQUILIBRICAL STRUCTURE IS A BALANCE OF AFFINITY PROPENSITIES, FLOWS PASSING THROUGH THE STRUCTURE, WITH THEIR PRESENCE, AND RESULTING AFFINITY PROPENSITIES WOULD DISBALANCE THE AFFINITY PROPENSITIES IN THE EQUILIBRICAL STRUCTURE.

Wherever there is the presence of a directed flow, there is an excess affinity propensity that would make other elementary particles orbit the flow to become a secondary flow, an inductive flow.

The elementary particles in the equilibrical structure, however, are not available to orbit the flow because they are in balance with the other elementary particles that make up the structure.

The directed flow of elementary particles will create a presence in the equilibrical structure, and the affinity propensity of the presence will affect the positioning of the elementary particles that make up the structure, causing them to move out of equilibrium.

As soon as the directed flow stops, however, the at-rest motion of the particles will move them right back to a position at which the affinity propensities of the structure will be in balance.

Thus, the equilibrical structure, made up of elementary particles with at-rest motion and affinity propensity can register the existence of directed flows of elementary particles by having the elementary particles in the structure affected by the affinity propensities of the directed flows move into disequilibrium.

And the movement will be proportional to the flows, with larger flows causing greater disbalances.

When the directed flows cease to move through the equilibrical structure, the elementary particles that make up the structure will move back into equilibrium.

The equilibrical structure can thus represent flows of elementary particles passing through it.

The equilibrical structure can therefore detect the flows of elementary particles released from the specific mixture of physical matter by the directed flows created by external noises.

By detecting these flows, chemical matter detects

external reality.

However, detecting variations in external reality is what the concept of the neural net does, and the neural net cannot compare.

How does the equilibrical structure that is going to form the basis for the mind compare?

It's going to compare reality with stored reality!

How is it going to do this?

If a pattern of flows is established in the equilibrical structure, then that pattern of flows, when established, disbalances the structure, and becomes the structure's norm.

Once the equilibrical structure is disbalanced by a pattern of flows, it cannot simultaneously be disbalanced by another pattern of flows.

IN ORDER TO COMPARE, THERE MUST BE SOMETHING TO COMPARE WITH. IF THE MIND IS GOING TO RECOGNIZE PHYSICAL REALITY, IT MUST HAVE A PICTURE OF PHYSICAL REALITY TO COMPARE IT WITH. THE EQUILIBRICAL STRUCTURE THEREFORE HAS TO HAVE THE CAPABILITY OF STORING A PICTURE OF PHYSICAL REALITY FOR RECALL.

Although there have been claims made to the contrary, the mind can only construct a single picture of reality at any single instance.

Thus, if the mind has been disbalanced by a pattern of flows that represents reality, and it attempts to accommodate a second pattern of flows that has been recalled to compare with the reality that the disbalanced particles represent, then whether or not the equilibrical structure can accommodate the second pattern of flows depends on the extent to which the second pattern of flows is identical to the first pattern of flows.

THE EQUILIBRICAL STRUCTURE CAN BE DISBALANCED BY RECALL OR REALITY. WHEN THE ONE THAT IS ATTEMPTING TO DISBALANCE THE EQUILIBRICAL STRUCTURE DOES NOT AGREE WITH THE ONE THAT IS ALREADY DISBALANCING IT, THE ELEMENTARY PARTICLES THAT MAKE UP THE STRUCTURE ARE MOMENTARILY FROZEN, AND THE INABILITY TO COMPARE GETS OUR ATTENTION.

If the pattern of flows disbalancing the equilibrical structure represent a tree, then a second set of flows that represent a pile of lumber is going to cause a severe conflict with the already disbalanced elementary particles making up the structure.

They are simply not going to be able to move to accommodate the second pattern.

Thus, if we are walking down the path to visit our favorite tree, and we have disbalanced the elementary particles in the equilibrical structure attached to our brains so that the disbalance represents our favorite tree, when we are confronted with a pile of lumber, the elementary particles that are already disbalanced in the equilibrical structure are not going to be able to accommodate the flows that represent the pile of lumber.

There is going to be a moment during which they have to come out of the disbalance that represents the tree and move into the disbalance that represents the lumber.

Because the disbalance is caused by directed flows of elementary particles, and because the equilibrical structure can only accommodate one pattern of flows at a time, there is going to be an excess of directed flows caused by the sight of the lumber.

Excess flows of elementary particles are directed into the subsystems so that an appropriate physical response can be achieved in the chemical matter.

The sight of the lumber gives you a feeling like you were hit in the solar plexus.

When the comparison process of the equilibrical structure of the mind fails, the failure sends a jolt of electricity to targeted areas of the body, making the organism aware of itself.

When we force a picture on someone that disagrees with the picture of self that person has, when we ridicule a person or otherwise belittle or make fun of that person's self, we are causing a conflict in the equilibrical structure of that person which immobilizes the structure and sends a jolt of current directly into that person's organs.

The comity of our laws generally only recognizes a physical assault represented by a slap or a kick as being actionable.

A slap on the face may well be less painful than a jolt of juice to the stomach, especially when the jolt is delivered over and over each time the affected person reconstructs the insult that creates the opposing picture of self.

This aspect of the operation of the equilibrical structure that makes up the mind needs emphasis.

The equilibrical structure developed to move the organism through physical reality by the process of comparing that physical reality with stored memories of physical reality so that the organism could be alerted to aspects of physical reality that might endanger it.

As long as reality accords with the state that exists in the structure, there is no attention getting shock. When there is a disagreement in the comparison process, the organism receives a shock designed to get its attention so that it can deal with the unexpected reality.

As soon as we obtain the ability to compare, however, we have the ability to comprehend what we compare.

We have the ability to associate the pile of lumber with a chopped down tree, and understand the processes involved in reducing it from its form as a tree to its form as lumber.

We can comprehend the difference between animate an inanimate matter.

But for the equilibrical structure to compare, it needs to get external reality converted to flows that can pass through the structure so that the flows can disbalance the elementary particles that make up the structure.

And once the equilibrical structure is disbalanced, it has to have a second source of flows, it has to have some way to recall that which it is going to compare with its picture of external reality.

And it has to have something to recall.

How do the external flows get into the equilibrical structure so that they provide an objective representation of external reality?

How does this equilibrical structure in turn recall pictures of external reality that are stored in memory so that they can be compared with the reconstruction of external reality in the equilibrical structure?

SO FAR, WE HAVE ONLY AN EQUILIBRICAL STRUCTURE IN WHICH THE COMPARISON OF REALITY CAN BE MADE WITH RECALLED REALITY. HOW DOES THE STRUCTURE OBTAIN THE FLOWS THAT REPRESENT EXTERNAL REALITY, HOW DOES IT STORE THEM, HOW DOES IT KNOW WHICH ONES TO RECALL FOR COMPARISON, AND HOW DOES IT RECALL THEM.

And how is it that the pictures recalled are pictures that are associated with the external reality constructed in the equilibrical structure?

Why, when we see a tree, we recall a tree to compare with a tree rather than a pile of lumber?

16. THE MECHANICAL MIND

Most modern models of the mind personify personal characteristics as its parts. Thus we have angels fighting devils, the self as ego fighting the dark desires arising out of the physical operations of the body, and generally, the conscious self fighting the unconscious self every step of the way.

Much of the problems we have in analyzing the mind is that the pictures constructed from the process of recall have the same effect on the physical subsystems as pictures constructed from reality, perhaps, because of idealization, more effect.

WE CAN ONLY ACT WHEN WE CAN CONSTRUCT A PICTURE OF OURSELVES ACTING. HOWEVER, WE CAN RECONSTRUCT PICTURES FROM BOTH REALITY AND THE REALITY WE RECALL. THEREFORE, OUR ACTS DO NOT NECESSARILY ACCORD WITH REALITY.

Our minds can provoke us into anger in physical reality, seduce us into passion.

We can provoke ourselves into rage by reconstructing a reality that no longer exists, and may never have, and seduce ourselves into passion by images that have no basis in reality.

However, our actions are limited by the pictures of reality that we can reconstruct in our mind because we cannot act without having a picture of ourselves doing so.

In like manner, our actions can exceed the proprieties of physical reality. We can do so by reconstructing a picture of physical reality that is not in harmony with actual physical reality. Acting in furtherance of that picture will be to the detriment of both physical reality and ourselves. It will be to the detriment of physical reality because disharmony throws physical reality into disbalance. It will be to our detriment because the acts become a part of physical reality, and therefore a part of the reality that we must live with.

We have to isolate the operation of the mind from the physical responses the mind raises in the organism which it guides through physical reality.

The physical responses that arise when the comparison process of the mind creates a conflict, an inability to compare, are absolutely necessary to the survival of the organism as it moves through reality.

The problem is, those responses are still operable when the mind is not being used to move the organism through reality.

The organism has physical reactions where the purpose of engaging in the comparison process is to construct a consistent picture of physical phenomena.

This occurs even though the process of contemplation does not provide a danger to the organism.

It is necessary to actually disengage the physical responses in order to isolate the mind's mechanical function so that we can see its parts, how they fit together, and how they result in the comparison process that leads to the contemplating that is producing the comprehension that we seek.

We have seen that an equilibrical structure can develop from the elementary particle with its two properties of at-rest motion and affinity propensity.

The elementary particles in equilibrium in the equilibrical structure become disbalanced to the extent of the presence of flows of elementary particles that pass through the structure.

HOW DOES EXTERNAL REALITY BECOME AN OBJECTIVE INTERNAL PICTURE OF REALITY IN THE MIND, SO THAT ALL MINDS RECONSTRUCT THE SAME EXTERNAL REALITY.

This disbalance then provides the basis for the comparison process that leads to comprehension.

How does external reality become a picture in the mind?

What is it about external reality that provides the information from which the mind can reconstruct a picture of that external reality?

There has to be a medium between external reality and the mind, and between the mind and recall, in which the information used to form a picture of reality, and reconstruct a picture of reality is encoded.

This medium is another equilibrical structure, an

encoded microstructure.

We don't have to look far in external reality to find out how the information about that external reality is conveyed.

All we have to do is ask the question, how do we know about external reality?

Other than the lesser senses of touch and hearing, we primarily know about external reality because streams of elementary particles are bouncing off of it. We have developed an eye that is sensitive to those streams of elementary particles.

ALL THE INFORMATION ABOUT EXTERNAL REALITY IS CONTAINED IN THE FLOWS OF ELEMENTARY PARTICLES, WITH THEIR AT-REST MOTION AND AFFINITY PROPENSITIES, THAT BOUNCE OFF THAT EXTERNAL REALITY.

So the information we have of external reality is the information contained in the streams that are bouncing off external reality.

The sun shines down on a rock, and somehow conveys the information about the shape of that rock to the eye.

We know that light expands over the surface of an expanding sphere and it does so inversely with the square of the distance from its source.

At every point in the expanding sphere, the presence of light is determined by its distance from its source.

Light from the same source at two different distances

from that source will each have a different presence.

Thus the expanding emission field that bounces off the rock has information associated with it at every point at which it hits the rock. This is because the light has a different presence at every point it strikes the rock.

It has a different presence because at every point it hits the rock, the rock is a different distance from the source of the light, in our case the sun.

The shape of the rock is therefore defined by its distance from the source of the emissions bouncing off of it. That distance is defined by the presence of the light that strikes every point of the rock.

We know that reflected light creates an additional center of expansion.

The light is striking the rock at different points. It therefore has a different presence at each point. Because it has a different presence at each point, it creates a new center of expansion at each point over which the light then expands.

What has happened by the process of reflection is that the information with respect to each position of each point on the rock in external reality is now fixed in that external reality by the presence of the flows that are bouncing off of it.

This situation occurs simply by the process of bouncing flows of elementary particles off of matter in reality. These flows bounce off the matter whether or not there is a detector to discern the differences in the presences of the resulting flows.

The movement of the flows between the source and an object in reality, and the subsequent reflection of those flows off the object creates an angle which establishes the presence of the reflected flows that exist in the reality around that object.

> **THE INFORMATION ABOUT EXTERNAL REALITY IS CONTAINED IN THE EXPANDING SPHERES OF ELECTROMAGNETIC EMISSIONS THAT BOUNCE OFF THAT PHYSICAL REALITY. THE PRESENCE OF THE FLOWS AT ANY POINT IN THE EXPANDING SPHERES IS DETERMINED BY ITS DISTANCE FROM THE FOCAL POINT, AND PRODUCES A PROPORTIONAL FIELD REPLACEMENT IN THE RECEPTORS OF THE EYE.**

They define the dimensions of that object in reality.

If we moved a single atom with a cloud of orbiting elementary particles around the reflected flows surrounding a rock, the ability of the nucleus of the atom to retain or lose the orbiting elementary particles would depend on the strength of the reflected flows because the atom's existence depends on the field in which it exists. The atom gives up its orbiting elementary particles as a result of field replacement.

If we had a really fine detector, we could measure the rate at which the elementary particles were released from their orbits. We could also measure the rate at which they were recaptured.

A series of atoms could be arranged to accomplish the same result. Instead of moving the atom with respect to the different fields, we could position a number of stationary atoms so that they could pick up the different flows and then compare them with one another for presence.

This series of atoms is actually the eye.

The eye is positioned with respect to the light bouncing off of objects in external reality. The light bouncing off objects in external reality contains information about those objects. This information results from the presence of the flows being dependent on the point at which the light bounced off the object.

The presence of the flows contains the information about the distance of all points from the eye that, when taken together with all of the information falling on the eye, allows the eye to fix the points of that object in space, and therefore fix the objects in reality.

The eye operates on the process of field replacement. The atoms of the eye detect the strength of the flows by their presence. They operate as detectors by releasing a proportional number of their elementary particles to provide the information about those flows.

The particles are released and form into an equilibrical structure that contains the information necessary to reconstruct the flows in the mind.

At each instance that the eye collects the information from the flows of elementary particles, field replacement occurs in the atoms of the eye. The field replacement produces a source of elementary particles that is proportional to the total presence of the flows that are replacing the affinity propensities of the atoms.

In addition, the combined flows have entered the eye to become a part of the electrical flows that drive the engine of the mind.

THE EYE ENCODES THE ELEMENTARY PARTICLES RELEASED IN THE PROCESS OF FIELD REPLACEMENT INTO AN EQUILIBRICAL MICROSTRUCTURE WHICH HAS A NET AFFINITY PROPENSITY EQUAL TO THE TOTAL OF THE FLOWS REPRESENTING EXTERNAL REALITY.

We now have an equilibrical structure produced from the elementary particles released in the process of field replacement and which represents flows of elementary particles in external reality.

We also have the elementary particles that make up the combined flows representing physical reality.

At any instance, the number of elementary particles released from the atoms of the eye and encoded into an equilibrical structure will equal the total of the flows that created the field replacement and are now entering the system.

The concept is that the charge of the flows of

elementary particles that caused the field replacement will equal the charge of the equilibrical structure formed by the particles released in the process of field replacement.

To delve briefly into the biology of the system, the tears are the source of the elementary particles that provide the atoms of the eyes with the elementary particles to replenish those lost from their orbiting clouds. The equilibrium is regained during the process of blinking which both provides a source of the elementary particles and momentarily cuts off the flows that strip the atoms of the elementary particles that are being replaced. The tears obtain the elementary particles from the sinuses, which, by moving water molecules with excess affinity propensity through them, regulate the electrical structure of the brain, eliminating excess elementary particles not needed in its operation.

The elementary particles that are released by the eye in the process of field replacement are captured into an equilibrical microstructure that represents the flows that released the particles.

We know that physical matter forms, because our environment is filled with it.

We know that physical matter forms into chemical matter as a result of its movement within the various emission fields it becomes subject to.

I suspect that spin is the key to chemical matter's ability to lock individual elementary particles into equilibrical structures.

The equilibrical microstructures that are formed in the eye have their at-rest motion and affinity propensity locking in each other.

The equilibrical microstructure forms in the eye. It is the result of the individual flows of elementary particles. The individual flows release a proportional number of elementary particles from the clouds orbiting the individual atoms. The released flows represent the original flows. The chemical process of the eye forms them into equilibrical microstructures. The excess affinity propensity of the resulting structure matches the excess affinity propensity of the combined flows that produced the field replacement that released them.

THE ELECTRICAL CHARGE OF THE MICROSTRUCTURE MATCHES THE ELECTRICAL PRESENCE OF THE COMBINED FLOWS OF ELEMENTARY PARTICLES ENTERING THE EYE. THIS FLOW CAPTURES THE MICROSTRUCTURE BY ITS AFFINITY PROPENSITY AND CARRIES IT INTO THE BRAIN FOR RECONVERSION INTO THE FLOWS THAT WILL DISBALANCE THE EQUILIBRICAL STRUCTURE.

We are referring to the equilibrical structure as a microstructure to distinguish it from the equilibrical structure in which the comparison process occurs.

The microstructure encodes physical reality, or reconstructions of physical reality.

At the point in the flow where this microstructure forms, the excess affinity propensity of the flow exactly equals the excess affinity propensity of the structure formed.

If we picture ourselves looking at a tree, then the flows from the various points of the trees are hitting the eye, passing through the atoms in the eye and combining into a single flow.

At the same time the flows pass through the atoms in the eye, the atoms release a proportional number of elementary particles representing the flows. These flows are encoded into a microstructure with an excess affinity propensity exactly equal to the excess affinity propensity of the combined flows that represent the tree.

THE EQUILIBRICAL STRUCTURE THAT MAKES UP THE MIND IS CONTINUOUSLY DISBALANCED BY THE FLOWS OF ELEMENTARY PARTICLES TRANSPORTED BY MICROCURRENTS THAT MATCH THE EXCESS AFFINITY PROPENSITIES OF THE MICROSTRUCTURES CARRYING THE ENCODED FLOWS.

The mind only needs about thirty reconstructions of reality a second to create a consistent picture if the number of frames of a motion picture it requires for the same purpose are any indication.

But these elementary particles are moving at the speed of light, and there are uncountable numbers of them.

It is best to visualize the process of looking at the tree as the process of the microstructures being formed and immediately attaching themselves to the new combined flow which whisks them off through the optic nerve into the chemical brain.

If you look steadily at the tree, then the microstructures, and the flow associated with them will stay at the same level, with the excess affinity propensities balanced at the point of similarity.

If you move your eye so that it sees nothing but the blue sky, then the resulting microstructure, and the associated flow will have a different excess affinity propensity.

Because we know that the flows of elementary particles are actually electrical flows, we can refer to them as a current.

Because it is the excess affinity propensity that is attracting and holding the microstructure that has an identical excess affinity propensity, we can say that the level of the current determines the microstructure that it can transport.

In the eye, the current always matches the microstructure because the two have the same source, the sum of the flows that are reflected off of objects in physical reality and which pass through the atoms of the eyes to form both the resulting flow and the microstructure representing the flows.

We have already seen what happens when the microstructure reaches the equilibrical structure that represents the mind. It is passed through chemical matter which causes the re-creation of the flows that the microstructure represents. The flows pass through the equilibrical structure, disbalancing the elementary particles that make it up.

The disbalance produces an objective representation of reality.

THE EYE CHEMICALLY CONVERTS THE FLOWS REPRESENTING EXTERNAL REALITY INTO A MICROSTRUCTURE. CURRENT FLOWS THAT MATCH THE EXCESS AFFINITY PROPENSITY OF THE MICROSTRUCTURE CARRY IT INTO THE MIND WHERE IT IS CHEMICALLY RECONVERTED BACK INTO THE FLOWS THAT PASS THROUGH THE EQUILIBRICAL STRUCTURE OF THE MIND.

Any organism that has a reasonably efficient eye, a reasonably efficient chemical converter and a reasonably balanced equilibrical structure would therefore have the flows which are converted from external reality to internal perception by the process of field replacement represent the proximate picture of reality by the disbalance of the elementary particles that make up the equilibrical structure of the mind.

Of course, the picture that results from going through the process of field replacement twice, first to form the microstructure and then to release the elementary particles that make it up, is not going to be an exact representation of external reality.

But it is sufficient to allow the organism to move within that reality in a manner that would extend its range of survivability.

When we look at the tree we recall a tree.

We had to recall a tree because if the tree we were seeing was the first time we had seen a tree, and we didn't have a tree to recall, then we wouldn't know that what we were looking at was a tree.

We wouldn't have the recognition that comparison provides.

Where did we get the microstructure of the tree we recalled?

As the flows pass through the equilibrical structure and disbalance the individual elementary particles, the elementary particles move back into position in response to their at-rest motion.

> **THE DISBALANCING OF THE ELEMENTARY PARTICLES THAT MAKE UP THE EQUILIBRICAL STRUCTURE IS ANALOGOUS TO THE FORMATION OF AN INDUCTIVE FLOW. THE REBALANCING OF THE STRUCTURE IS ANALOGOUS TO A COLLAPSING INDUCTIVE FIELD. A COLLAPSING INDUCTIVE FIELD ITSELF ESTABLISHES FLOWS OF ELEMENTARY PARTICLES.**

In the real world, the creation of a flow of electricity induces a secondary flow, the inductive flow. A collapsing inductive flow in turn induces a primary flow of electricity.

The flows disbalance the elementary particles in the equilibrical structure, and when the particles move back into balance they themselves induce a flow that is identical to the flows that created them.

The result is that the equilibrical structure re-encodes the flows into both a single flow and a microstructure in a manner similar to the way the eye operates.

The resulting microstructure will now be thrice removed from external reality, and therefore will depict that external reality to a lesser degree than the original reconstruction of that reality in the equilibrical structure.

It will still be sufficient to allow comparison and comprehension of that reality when it is recalled.

THE FLOWS CREATED WHEN THE ELEMENTARY PARTICLES IN THE EQUILIBRICAL STRUCTURE MOVE BACK INTO POSITION ARE CHEMICALLY RE-ENCODED INTO A MICROSTRUCTURE WITH AN AFFINITY PROPENSITY THAT MATCHES THE CHARGE OF THE COMBINED FLOWS, THE RESULTANT CURRENT CARRYING THE MICROSTRUCTURE INTO THE BRAIN FOR STORAGE.

The excess affinity propensity of the microstructure again matches the current flow, and the microstructure is whisked to the neurons, where it is stored for future recall.

The current flow generated by the mind is the filing

system in the memory storehouse of the chemical brain. The chemical brain captures the encoded equilibrical microstructures of memory until a current flow passes that can dislodge them from their perch.

Of course, if you are looking steadily at the tree, the steady current that is created will continue to store and recall the same microstructures. You can actually perfect the external image to the extent that you are recalling the same thing you are looking at.

In that case, the object will lose its meaning.

Without comparison, comprehension ceases.

When you look at the sky, the current flow changes completely, the picture of the tree dissolving in the equilibrical structure.

If you want to retrieve the picture of the tree you have to alter the current flow to a level that will once again pick up the microstructure that represents the tree, carry it to be decoded with the resulting flows sent through the equilibrical structure to once again form a tree.

This is the process of comparison.

The input into the equilibrical structure that makes up the mind comes from two sources, is simultaneous and can be varied with respect to the importance of the source.

You can look at the tree and consciously alter the flows that are coursing through your chemical brain, turning the branches into a shroud, the dew drops into diamonds simply by painting with the brush of recall rather than the brush of reality.

The equilibrical structure is driven by the microstructures created in the eye. As long as the eyes are open, the microstructures are being carried to the brain and converted to the individual flows which represent external reality and which stream through the structure, disbalancing it in a specific pattern that represents that reality.

However, because flows from microstructures that are recalled reality are constantly streaming into the equilibrical structure, and those microstructures are themselves re-encoded, three steps removed from reality, the external reality that is being reconstructed is being modified by the imperfect re-encoding of that reality, so that what is being constructed in the equilibrical structure in the first place is altered by recall.

MICROSTRUCTURES ARE RECALLABLE BY CURRENT LEVELS THAT MATCH THEIR EXCESS AFFINITY PROPENSITIES. THE PICTURES FORMED IN THE EQUILIBRICAL STRUCTURE DETERMINE THE CURRENT LEVEL. A CURRENT LEVEL CREATED BY A PICTURE OF A TREE IN THE MIND WILL PICK UP A MICROSTRUCTURE FROM ANYWHERE IN THE BRAIN THAT IS AN ENCODED PICTURE OF A TREE BECAUSE ITS EXCESS AFFINITY PROPENSITY MATCHES THE CURRENT FLOW CREATED BY THE PICTURE OF THE TREE IN THE MIND.

We never do see the reality that exists exactly as it appears in front of our eyes unless we look at it long enough so that recall exactly matches what the eye is encoding.

And when that happens, we lose comprehension!

The system was developed to move the organism through reality.

It was designed to convey information to the organism about external reality that allowed the organism to avoid the pitfalls of that reality. If we are moving through external reality, as long as the system provides recall microstructures that match the microstructures that are being created by the engine of external reality, there is no message sent to our subsystems. We can drive down the road without reacting to everything in the environment as long as the recall process can match the incoming microstructures with recall microstructures.

THE INABILITY OF THE EQUILIBRICAL STRUCTURE TO FORM MORE THAN ONE PICTURE OF REALITY FORCES IT TO MODIFY THAT REALITY TO CONFORM TO THE IMPERFECT RECALL PROCESS. THE ABILITY TO ALTER THE PICTURE OF REALITY, HOWEVER, IS WHAT ALLOWS THE MIND TO CREATE A REALITY THAT DOESN'T EXIST, CONSTRUCT IT IN REALITY, AND THEREBY EXTEND ITS RANGE OF SURVIVAB ILITY.

But if external reality starts to feed a picture of a car coming directly toward us, then the elementary particles that make up the equilibrical structure cannot accommodate the two pictures. Their physical position of disbalance in the structure cannot accommodate the new picture that is attempting to be constructed, and if they could, there would be no recall microstructures to agree with it.

The equilibrical structure ceases to accommodate the conflicting flows. The smooth process of the system, with flows coming from the eyes and joining the flows circulating through the chemical brain, comes to an abrupt halt because when the structure ceases to operate, the re-encoding process ceases.

The electricity does what it was supposed to do.

It enters the subsystems and shocks various organs so that the organism becomes aware very quickly of the changed conditions in external reality.

The system, having performed its function, then returns to normal in an attempt to deal with the situation, either successfully or unsuccessfully.

The system works to move an organism successfully through external reality.

However, the very fact that it cannot match external reality with recalled reality results in the comprehension that allows the creation of worlds with varying degrees of concordance with external reality.

Once there is an internal reality, and the ability to create worlds that don't exist in external reality, there is the ability to create a picture of external reality and alter it so that it will better accommodate the organism.

> **THE SCIENTIFIC METHOD IS A PROCESS OF TRIAL AND ERROR. WE CONSTRUCT A PICTURE OF REALITY AND TEST IT BY ATTEMPTING TO CREATE THE PICTURE IN REALITY. IF IT DOESN'T WORK IN REALITY, WE BUILD ANOTHER PICTURE OF REALITY UNTIL WE FIND ONE THAT WORKS. THE METHOD PERPETUATES NEGATIVES BECAUSE NEGATIVES CANNOT BE DISPROVEN. THE PROBLEM WITH NEGATIVES IS THAT IF YOU ACCEPT THEM, YOU WILL NEVER FORGE PICTURES THAT WILL ALLOW YOU TO ACT TO OVERCOME THE NEGATIVES. GRAVITY AS A PROPERTY OF MASS IS THE UNIVERSAL NEGATIVE, IT BEING PHYSICALLY IMPOSSIBLE TO OVERCOME A PROPERTY.**

Because acts can only be taken if the organism is capable of constructing a picture of itself acting, the organism can act only in furtherance of the picture of altered reality that it has constructed.

This is basically the scientific method.

We create a picture of external reality that we believe to be true, and then we act in accordance with that picture, using the results of our acts to determine the validity of the picture.

Because individual acts do not validate the totality of external reality, we have to construct a picture of external reality that is consistent, and hope that the consistency validates it, knowing that it can only validate it to the extent that it agrees with the facts.

And, of course, if our picture of external reality is negative, contains prohibitions to actions such as you can't go faster than the speed of light, you can't create a physical system that will deal with gravity because gravity is a property of and proportional to the physical system, you can't go to the moon, then we will not form the pictures that will allow us to act.

We will never look past the key that turns in the ignition.

We will never get anywhere!

A system that evolved to allow an organism to move successfully through external reality by the very nature of its imperfection became a structure that could create worlds, and analyze them so that it could comprehend the world in which it found itself.

The awareness that resulted from the equilibrical structure, the awareness that is necessary to move an organism through physical reality, created a picture of the organism in that external reality that gives rise to the "I" of the organism.

Assuming that the equilibrical structure reconstructs reality at the rate of thirty pictures a second, it is re-encoding external reality in thirty microstructures a second.

Each one of those microstructures contains a picture of external reality.

Part of that external reality is the organism in which

the system is operating, so that sooner or later the recall microstructures start to reconstruct a picture of that organism with the pictures of external reality.

Eventually, the "I" becomes the constant in all recall.

EACH MICROSTRUCTURE ENCODED AT THE EYE AND FORMED IN THE EQUILIBRICAL STRUCTURE CONTAINS THE DIMENSIONS OF EXTERNAL REALITY THAT INCLUDES THE ORGANISM DOING THE ENCODING. EACH PICTURE RE-ENCODED AND STORED IN THE CHEMICAL BRAIN CONTAINS A PICTURE OF THE ORGANISM DOING THE ENCODING. THUS, EACH ORGANISM HAS ONE ELEMENT IN COMMON IN EVERY MICROSTRUCTURE IT PRODUCES AND STORES: "I"!

The current level of the re-encoded microstructures that include the organism's interaction with external reality makes the recall of those microstructures dependent on being able or willing to generate that specific current flow.

If microstructures aren't recallable, then they will never be constructable in the equilibrical structure.

In one form of sleep, where the system has been shut down, its engine silenced by the closed eyes, there is no recall, and thus the series of microstructures that contain the "I" are in storage.

There is no "I".

But what is sleep, and what are dreams?

As the organism moves through external reality, the changing external reality continually alters the disbalance of the elementary particles that make up the equilibrical structure.

Instead of the equilibrical structure regaining its equilibrium, it continues in disbalance, its elementary particles moving from one position to another without moving back into the position that represents the structure in equilibrium.

WE MAY NOT BE ABLE TO RECALL SOMETHING BECAUSE WE ARE NOT WILLING OR ABLE TO GENERATE THE LEVEL OF CURRENT NECESSARY TO PICK UP THE MICROSTRUCTURES FROM THE CHEMICAL BRAIN. IF WE ENCODED THE MICROSTRUCTURES WITH THE HELP OF ALCOHOL, OR AS A RESULT OF BEING SEVERELY BEATEN, THEY MAY BE INACCESSIBLE BY THE CURRENTS GENERATED WHEN THE MIND IS OPERATING IN A NORMAL FASHION.

The more in disbalance the elementary particles are when they move to accommodate the flows that represent external reality, the less they will be able to move to accommodate those flows.

As their ability to accommodate the flows that represent external reality decreases, the picture of external reality that the equilibrical structure forms grows further and further away from that external reality, and the recall microstructures that match the imperfect picture create increasingly imperfect representations of external reality to be re-encoded.

Because we act on the basis of the pictures of external reality we construct, our actions increasingly do not reflect external reality.

Our muscles may well have the ability to continue to operate, but they are operating in response to inaccurate representations of external reality.

The equilibrical structure needs an opportunity to regain its equilibrium.

We shut our eyes to stop the flows that drive the engine of the mind.

However, the elementary particles that make up the equilibrical structure are still in disbalance.

They have to get from their positions of disbalance to their positions of balance.

As they do so, they act like a collapsing inductive field, creating flows without regard to the existence of incoming flows from external reality.

The induced flows are encoded and establish a current level that moves into the chemical brain and retrieves recallable microstructures that match the current flow.

These recallable microstructures move into the equilibrical structure and create pictures that match the movement of the elementary particles' movement. The pictures might be similar to pictures constructed from reality, or they might be totally random.

Wherever they come from, they result in the recall

process. Because the recall process is reconstructed filaments of reality, the eyes move back and forth, extracting elementary particles to fuel the process. The process continues until the elementary particles that make up the equilibrical structure regain their balance.

At which point the dreaming stops!

The system shuts down until the organism becomes aware of reality once again and the engine starts up, cycling the "I" into the equilibrical structure to once again deal with the demands of that reality.

IF WE CONCENTRATE ON AN OBJECT IN REALITY, IF WE ARE STUNNED BY A REALITY THAT OPPOSES OUR PICTURE OF REALITY, THE ELEMENTARY PARTICLES THAT MAKE UP THE EQUILIBRICAL STRUCTURE BECOME STATIONARY, AND THERE IS NO COMPREHENSION, NO COMPARISON, NO ENCODING OF REALITY, NOTHING BEING RECALLED. IF WE ARE ASLEEP, WITH THE PARTICLES IN PERFECT EQUILIBRIUM, THERE IS NO RECALL, NO MICROSTRUCTURES MOVING THROUGH THE STRUCTURE, NO "I". IT IS ONLY WHEN THE SENSES BEGIN TO CONVERT REALITY INTO FLOWS OF ELEMENTARY PARTICLES THAT THESE FLOWS RESTART THE ENGINE. THE FLOWS ARE WHAT DRIVES IT, DISBALANCING THE PARTICLES, CAUSING THEM TO REBALANCE, TO ENCODE, TO RECALL, IN SHORT TO ENTER ONCE AGAIN INTO THE STATE OF AWARENESS.

17. THE AGONY OF INACTION: A DEFECT IN COMPREHENSION

The equilibrical structure is attached to the chemical brain by the stabilization of excess affinity propensities between the structure and the brain.

The structure generates the mind, the perception of reality, and the comparison of that reality with stored reality.

It does so by constructing a picture of reality generated by the flows of elementary particles that actually bounce off of that reality.

The picture constructed from reality is re-encoded at a specific current level which both serves as the storage marker of the picture, and the recall marker.

THE CURRENT LEVEL FLOWING THROUGH THE BRAIN IS ITS FILING SYSTEM. PICTURES OF REALITY ARE FORMED AT THE SAME CURRENT LEVEL AT WHICH THEY ARE RE-ENCODED AND STORED. WHEN A PICTURE OF A TREE IS FORMED IN THE MIND, THE MIND GENERATES A CURRENT FLOW THAT RECALLS THE PICTURE OF A TREE STORED AT THE SAME CURRENT LEVEL. WE USE OUR MINDS BY LEARNING HOW TO MANIPULATE THE LEVEL OF THESE FLOWS.

As the pictures of reality are constructed in the equilibrical structure, the re-encoding process generates a current that searches the storehouse of the chemical brain for microstructures stored at similar current flows.

As the current flows move throughout the brain, it makes little difference where the microstructures dealing with the pictures of recallable reality are stored.

It is the current flow itself which is the filing system!

External reality is the engine that drives the construction of pictures of reality in the equilibrical structure, and it is the picture that is constructed that dictates the level of current flow, which is identical for both what is re-encoded and what is recalled.

The system has to run smoothly in order for it to provide the shock that is designed to get the organism's attention when the picture of external reality all of a sudden disagrees with the recall process.

The difference in the picture of what reality should be and what reality is creates a conflict, and it is this conflict that allows the organism to not act for as long as it takes to sort out its position in reality so that it can continue to move safely in the changed reality.

The organism recognizes the existence of the conflict when the equilibrical structure ceases to function smoothly. The current flows that went into the operation of the equilibrical structure are being redirected into the organism itself, destabilizing the chemical matter in the organism through field replacement so that the organism will take action to rebalance the disequilibrium.

If the car starts to slip on the ice, careening out of

control, our mind spins as it ceases to process reality, and the flows that went into its operation enter the body causing the stomach muscles to contract, the palms to sweat, the bowels to move. We frantically take action to regain control, the actions themselves being a method of regaining the equilibrium in the subsystems, and thus getting the mind to function once again.

THE MIND WORKS AS A DETECTOR BECAUSE IT IS INCAPABLE OF FORMING MORE THAN ONE PICTURE AT A TIME. WHEN A PICTURE PRESENTS ITSELF TO THE MIND THAT DOESN'T MATCH THE PICTURE IN THE MIND, THE MIND CANNOT FORM A PICTURE OF BOTH. THE MIND STOPS FOR A BEAT AND THE CURRENTS THAT NORMALLY OPERATE THE MIND INVADE THE BODY, REPLACING EXISTING FIELDS IN THE BODY THAT AWAKENS THE MIND TO THE NEED TO REFOCUS ITS ATTENTION ON THE CHANGED REALITY.

The whole purpose of the development of the equilibrical structure is to generate action.

It does so by being sensitive to changes in physical reality. The changes cause a conflict which, if not handled by existing experience, the recall of microstructures that tell you how to turn the wheel, redirect the electrical flow from the mind's operations to the body, disbalancing the subsystems.

The disbalance is perceived as discomfort, something not to be desired, and the organism takes action to remove itself from the stress caused by the conflict.

The mechanism is predictable and unrelenting.

Conflicts which can't be resolved by recall open the switch to the subsystems as simply as turning on a light heats up the light bulb.

Conflicts cause stress in the chemical matter that makes up the organism by allowing electrical flows into the subsystems that make up the organism, releasing the elementary particles that hold the atoms of that matter together to create additional stress in the organism.

The purpose of the stress is to make the organism aware of the conflict so that it can take action to avoid the danger created by the conflict.

This causes the organism to cease to act in the manner that created the conflict.

On this level the mechanism of the mind is similar to the mechanism of chemical matter that releases its orbiting elementary particles in response to electrical flows generated by sound, taste, touch or smell.

The reaction of chemical matter is sufficient to provide the method of stressing the organism so that it can be aware of the external stimuli.

However, chemical matter could never respond to the diverse flows of elementary particles that make up physical reality, the sunlight that defines physical matter in objective reality in a manner that allows an organism to move successfully through that external reality.

Just as physical matter developed chemical matter as a result of the environment changing with respect to the physical matter, when the chemical matter had to move within the environment, it developed equilibrical matter to allow it to interact with the environment it had to move in response to.

Its original purpose was no different from the chemical matter that served as detectors of external reality, inducing the organism to move so as to diminish the disbalance in order to regain its equilibrium.

EQUILIBRICAL MATTER AS AN OUTGROWTH OF CHEMICAL MATTER IS AS CERTAIN AS CHEMICAL MATTER IS AN OUTGROWTH OF PHYSICAL MATTER.

When we realize that the mind, the equilibrical matter evolved as a detector to allow the organism to avoid conditions in physical reality that would be harmful, then we can start to analyze its responses when the mind is used to analyze physical reality independent of the need to react within the environment to protect itself.

The purpose of analyzing the physical environment is identical to having the ability to move safely within that environment. Both extend our range of survivability with respect to the environment.

When we feel an increase in temperature on our skin, we withdraw our arm. The physical response creates helpful behavior.

When we see a car coming directly at us, we get off the road. The physical response creates helpful behavior.

When we construct a consistent picture of physical phenomena, we can create temperature controlled houses. The physical response creates useful behavior.

THE MIND RECALLS REALITY SO THAT CONFLICTS WITH EXTERNAL REALITY CAN SHOCK THE SYSTEM, ALERTING THE MIND TO CHANGING REALITY SO THAT IT CAN EXTEND ITS RANGE OF SURVIVABILITY IN THAT PHYSICAL REALITY. HOWEVER, THE MIND CAN CREATE REALITIES WITHOUT MOVING THROUGH EXTERNAL REALITY, FORM IMAGES THAT AFFECT THE BODY WITHOUT REFERENCE TO PHYSICAL REALITY.

Where does the shock to the system come from that drives us to build houses?

If the equilibrical structure is designed to shock the system in response to conflict, then it operates on the absence of conflict.

The system is at rest when there is no conflict.

The condition in which there is no conflict is a condition of agreement.

When there is agreement, there has to be two conditions in which there is agreement.

In the equilibrical structure, the two conditions are the flows produced by external reality and the flows produced by recall.

As long as the flows from these two sources are in agreement, there is no shock to the system.

Just as we can't relax when someone is sticking pins in our flesh, so we can't relax when there is a disagreement between what we can call external reality and recalled reality.

As we drive down the road, external reality is driving recall reality so that the two inputs are in agreement.

When external reality changes, the change is instantly detectable because it no longer agrees with recall reality.

THE RECALL FUNCTION ESTABLISHES NORMAL MENTAL ACTIVITY IN AGREEMENT, A BASELINE SO THAT CHANGES IN THAT NORMAL LEVEL OF ACTIVITY SHOCKS THE SYSTEM.

The conflict alerts the organism.

If we are sitting out in the cold, and we can recall being warm, then our recall reality is going to disagree with external reality.

The external reality exists. The recall reality is constantly recallable because its relationship to external reality has caused it to be encoded at a same or similar current level.

Cold makes us recall warmth!

The conflict drives the organism to alter external reality so that it agrees with recall reality.

The ability to change the external reality to match recall reality will depend on the development of technology. The development of technology in turn depends on both trial and error, pawing at the ground until you learn to burrow, and developing a consistent enough picture of reality so that you can pursue the possible.

Why is it that we end up limiting the possible by accepting concepts that delimit reality and proscribe our development?

OUR BODIES CAN FORCE OUR MINDS TO CREATE A PICTURE OF REALITY THAT DOESN'T EXIST, A SHELTER AGAINST THE COLD. THE RESULTING CONFLICT WILL DRIVE US CEASELESSLY TO ALTER REALITY TO ACCOMMODATE RECALLED REALITY.

While we don't control external reality, we have absolute control over recall reality.

When Copernicus suggested that the Earth was not the center of the universe, but actually traveled around the sun, the recall reality that had developed as the shared reality, that the sun traveled around the Earth, agreed with the observed reality, the sun moving in the sky.

Shared reality, because it reinforces the basic operation of the equilibrical structure to be balanced in agreement, is a powerful tool for creating an agreement between recall reality and external reality where external reality is not directly measurable. We ignore facts that disagree with our concepts and find facts that do agree.

What did Copernicus' suggestion that the Earth traveled around the sun create in the equilibrical structure of those whose recall reality viewed the Earth as the center of the universe?

It created a conflict.

What is the effect of a conflict in the equilibrical structure?

Pain.

Not imagined pain.

Real, honest pain, the result of the electrical discharge that the conflict sends into the body, flooding its subsystems with elementary particles moving at their at-rest speed, creating field replacing affinity propensities that disrupt the systems, tear them apart.

WHEN THERE IS NO OBJECTIVE REALITY, THE BASIC OPERATION OF THE MIND, ITS TENDENCY TO ESTABLISH A BASE LINE ROOTED IN AGREEMENT, CAUSES SHARED REALITY TO BECOME REALITY.

The equilibrical structure produces conflicts to warn the organism about a danger in physical reality.

Here there is no danger in physical reality.

The fact that the Earth travels around the sun may make a difference to us today, now that we know that the Earth isn't forever, and our very survival ultimately rests with our ability to perceive physical reality accurately.

What in the world difference did it make to someone sitting in a monastery five hundred years ago?

When our shared reality is opposed by another proposed reality, the conflict is just as real as the conflict that arises when another car heads straight for us on the highway.

The pain is worse because the conflict is with a reality that has been accepted as faith.

There was no way to prove one way or the other whether the sun went around the Earth or whether the Earth went around the sun.

DISAGREEMENTS WITH SHARED REALITY, SUCH AS THE REALITY OF THE MASS/GRAVITY CONCEPT, CAUSE CONFLICTS THAT RESULT IN ACTUAL PHYSICAL PAIN TO THOSE SHARING THE REALITY.

Even today, the proof is elusive. We can never encompass the solar system with our measuring devices.

However, there is very little in physical reality that the concept of the sun going around the Earth would accord with other than the fact that it moves across the day sky.

The point is, however, that in the year fifteen hundred, both were nothing more than concepts.

When we mistake concepts for reality, we create a belief that is unsupported by fact.

Beliefs are never far from our mind.

When those beliefs are challenged, then the challenge forces a picture of reality into the equilibrical structure that is not in accord with recall reality.

The conflict floods the organism's subsystems with elementary particles with painful current flows.

There is no action in reality that will resolve the conflict where the reality being conflicted is a shared reality rather than an actual physical reality.

Every time we form a picture of our cherished belief, the currents generated by the picture attempt to form a picture of the opposing concept.

We can't escape it.

We bang our fist, the muscular activity dissipating the elementary particles.

WHEN THE CONFLICT IS BETWEEN RECALL REALITY AND REALITY, THE CONCEPT WILL BE SHORT LIVED BECAUSE REALITY WILL ELIMINATE THE CONFLICTING RECALL REALITY (OR THE ORGANISM CLINGING TO THE RECALL REALITY). WHEN THE CONFLICT IS BETWEEN TWO CONCEPTS, TWO PICTURES OF RECALLED REALITY, ONE WILL ALWAYS EVOKE THE OTHER, AND THE CONFLICT WILL CONTINUE UNTIL RESOLVED.

We scream, the muscular activity dissipating the elementary particles, the noise producing electrical flows from the ear that place the particles in useful activity, we cry, our tears removing large quantities of the particles.

We burn the challenger at the stake which is one way

to eliminate the conflict altogether.

Another way to eliminate the conflict is to ridicule it.

Ridicule removes it from consideration. It eliminates the possibility of conflict by refusing to allow the equilibrical structure to create a picture of reality that would include the concept.

CONCEPTS THAT OPPOSE BELIEFS, SHARED REALITIES, ARE HARDER TO DEAL WITH BECAUSE THERE IS NO WAY TO ACCOMMODATE A CONFLICT IN BELIEFS. RETREATING FROM REALITY WILL NOT AVOID THE CONFLICT BECAUSE THE CONFLICT IS RAISED BY MERE THOUGHT. JUST AS BEING COLD RAISES THE THOUGHT OF BEING WARM, TRYING TO REAFFIRM A BELIEF CREATES REAL PAIN BECAUSE IT RAISES THE OPPOSING BELIEF, CREATING AN ENDLESS CYCLE OF CONFLICT.

Ridicule is indeed a luxury.

After all, the whole purpose of the equilibrical structure is to warn of danger in the environment. It is designed to move the organism into balance, to get it out of the pain created by the conflict so that there is no conflict.

Why shouldn't we luxuriate in the ridicule that will eliminate the need to consider concepts that oppose our picture of recall reality, the picture we accept as external reality and share with those we exist with.

It took Galileo's telescopes to show the moons of Jupiter traveling around Jupiter, rather than Jupiter traveling around the moons, and create a picture of reality that produced a conflict with the recall reality of the sun traveling around the Earth.

The conflict that the picture of reality created opposed the concept of the sun traveling around the Earth so that the only way to get rid of the conflict, and the resulting pain, was to alter the recall reality to conform to reality.

BECAUSE AFFIRMING SHARED CONCEPTS RECALLS OPPOSING CONCEPTS, AND BECAUSE SHARED CONCEPTS ARE ACTUALLY PART OF THE CONTINUALLY RECALLABLE "I" WITH AN OBJECTIVE REALITY ALL OF THEIR OWN, THE ONLY WAY TO AVOID A SORT OF PERPETUAL PAINFUL RAGE RESULTING FROM A CONTINUING CONFLICT, ASIDE FROM HANGING THE AUTHOR OF THE CONFLICT, IS TO RIDICULE THE OPPOSING CONCEPT. RIDICULE ELIMINATES OPPOSING CONCEPTS FROM CONSIDERATION, ELIMINATING THE POSSIBILITY OF PAIN.

What difference does it all make so long as the equilibrical structure ends up in balance? If ridicule relieves the subsystems from stress, why not ridicule away?

At a certain point in human development it may have been necessary to have the pain produced by the conflict drive comprehension to come up with ways to alter external reality so as to eliminate the conflict.

We all need food and protection against the

environment.

But there is a point at which the pain produced by the conflict becomes counterproductive.

That point is reached when the danger presented by physical reality is so remote that it fails to create a conflict.

Without a conflict, there is no pain, and no impetus to alter external reality.

It is easier in such situations to resolve the opposing conflicts by consigning them to ridicule so that the conflicts can be avoided in the first place.

The problem with this method, however, is that it doesn't recognize that the pain producing conflict is a defect in the operation of equilibrical matter when the purpose is not to resolve conflicts between external reality and recall reality, but rather to produce a recall reality that agrees with the external reality.

CHALLENGING THE SHARED REALITY OF MASS/GRAVITY IS NOT WORTH THE PAIN IT CREATES SO LONG AS ACCEPTING IT DOES NOT AFFECT OUR EXISTENCE IN REALITY. HOWEVER, IT IS ONLY JUST BEGINNING TO DAWN ON US THAT IF WE DON'T GET OFF OF THIS PLANET, WE ARE DOOMED TO DIE ON IT. THIS ACTUAL REALITY WILL EVENTUALLY FORCE US TO OVERCOME THE PAIN INVOLVED IN ACCEPTING THAT WE CAN DIRECTLY MANIPULATE THE ATTRACTIVE FORCE, THAT IT IS SOMETHING OTHER THAN A PROPERTY, SOMETHING WITH MEANING THAT IS UNDERSTANDABLE.

When we move beyond what is measurable, with what causes a rock to fall to the Earth, where does light come from, how our mind works, the only reality that exists, other than simple existence, is recall reality.

We can't limit our inquiry to what is measurable, and we certainly can't define our inquiry by what we think is measurable.

We have to use concepts as tools in the inquiry.

We have to create concepts, and then we have to evaluate them for their consistency with the physical reality they are created to describe and then with each other.

BECAUSE THE METHODS OF TRIAL AND ERROR ARE OF NO USE IN EVALUATING CONCEPTS, WHEN WE PASS THE POINT OF PROGRESS PROVIDED BY THAT PROCESS, WE HAVE TO DEVISE ALTERNATE METHODS TO GOVERN OUR MENTAL PROCESSES, TO ENSURE THEIR CLARITY IN REFLECTING PHYSICAL REALITY, AND THUS OUR FUTURE AS A SPECIES.

The physical warning system that evolved with the equilibrical structure to allow that structure to move the organism through physical reality does not work when it is applied to the structure when we are using it to come up with a consistent picture of physical reality.

A warning system is not appropriate when the equilibrical structure is being used to evaluate opposing concepts.

When it is used in the process of comprehending reality, it is a defect!

Concepts do not exist in physical reality.

We create them to comprehend physical reality.

They are only valid in so far as they explain physical reality.

If we allow the concepts we create to comprehend physical reality to replace the physical reality we are trying to comprehend, as we have done with the mass/gravity concept, then we will choose our concepts so that they will not conflict with our created concepts, moving further and further away from a description of physical reality in the process.

We cannot choose our concepts of physical reality on the basis of avoiding pain.

We have to disconnect the defect, server the connection between the conflict and the pain by recognizing that concepts are not physical reality, but merely a tool we use to explain physical reality.

If we don't, we will resolve the conflict by being a part of the planet as it slowly evolves into the condition of the moon.

Our progression from physical matter to chemical matter to equilibrical matter will reverse itself, with the equilibrical matter disconnecting from the chemical matter as it ceases to react in a changing environment.

We will have evolved to comprehend the physical universe, but, failing to overcome the defects of our evolution, we will have failed, and will simply rejoin the physical universe from which we evolved.

We will cease to comprehend.

18. THE PURPOSE OF COMPREHENSION

When we say that gravity is a property of matter, we are saying that the attractive force is not comprehensible.

The reason that we say the attractive force is incomprehensible is that we don't comprehend it.

We have not developed the concepts which will allow us to explain it as an operational part of physical reality.

There is no question, however, that the attractive force exists.

Therefore, because it is a part of reality, some sort of explanation has to be made for it, something that will keep the mind from being in a constant state of conflict, with the resulting undesirable physical consequences.

HOW WE APPROACH THE CONCEPTS WE USE TO ANALYZE PHYSICAL REALITY WILL DEPEND ON WHETHER WE WANT TO FEEL GOOD FOR A LIFETIME OR WHETHER WE WANT TO SURVIVE AS A LIFE FORM, A CHEMICAL STRUCTURE THAT PRODUCES EQUILIBRICAL MATTER.

The thought that any explanation is better than no explanation is valid if the purpose of the explanation is to quell the physical pain that results from not being able to comprehend reality.

If all we are attempting to do is quell our unsettled physical beings in the face of a comprehension that eludes us, then any explanation is indeed a good explanation.

If we are attempting to survive, then the only explanation that can satisfy us is an explanation that explains.

And explains in a manner that allows us to see how physical reality works.

Many of the tools modern science brings to the analysis of physical phenomena are not tools at all, but closures, explanations which, rather than explain, simply prevent the mental conflicts that cause physical discomfort.

Something being a property of something else is one way to accomplish this closure.

Thus, mass is meaningless without gravity and gravity is meaningless without mass, and as a result, both are meaningless.

$E=mc^2$ is a glorious conception if it is meant to convey the concept that the elementary particles that make up matter also make up the emissions from that matter we measure as energy.

We, however, take it to mean something else, and assume that it is actually an equation, that we can actually measure the amount of energy that comes out of matter, that we can actually measure the matter out of which it is coming. We actually believe that the square of the rate that light travels is related to the amount of energy that matter contains.

It is impossible to determine the number of elementary particles that make up matter.

It is impossible to determine the number of elementary particles that make up light.

Physical laws sterilize comprehension.

Who can argue with a law that states that an object will move in a straight line unless a force acts upon it to change its movement?

The problem is, the situation doesn't exist in reality because there is no object existing that does not have a force acting upon it, and objects come to rest with respect to the forces acting on them.

They may move perpetually in someone's recall reality, but they don't in physical reality.

We may well look at the sun as it burns in the sky, and come to the conclusion that laws dictate that everything runs down. We don't look at what is happening as the release of energy, the increase in the overall activity of the universe, but as the use of energy, its destruction, and this despite the fact that it is in total disagreement with the idea that is at the basis of universal operation, that objects can move in a straight line forever.

Of course, when we find that the solar system is not running down, we can create a new law, that matter can never be created nor destroyed and twist it around to apply to a solar system by saying that in a closed system energy is conserved. Thus, if the Earth speeds up a little, Mars, or perhaps three asteroids and a small moon slow down.

Modern science rests on a system of laws and principles, axioms and conclusions that lie beyond the realm of analysis by the scientific method because the scientific method can only disprove facts. It cannot prove hypotheses.

When Newton and his cronies (see Appendix) were called on to prove the occult concept that matter had some sort of attractive force associated with it, they came up with an amalgam of parlor tricks from osmosis to static electricity to convince the gullible.

BY BELIEVING THAT THE SCIENTIFIC METHOD VERIFIES HYPOTHESES, WE END UP BELIEVING A BUNCH OF CONCLUSIONS THAT HAVE NO BASIS IN REALITY, AND USE THE NON-FACTS THEY PRODUCE TO CREATE HYPOTHESES THAT MOVE US ONE STEP FURTHER INTO UNREALITY.

When this didn't work for those wishing to engage in rational discourse, when reasonable people kept pointing out that the idea was neither provable nor disprovable, the response was that the force was obvious, only the cause was occult, a notion that is carried into modern times by the statement that as long as we can measure the effects of gravity, we don't have to know its cause.

Why something falls is a philosophical question.

But the use of the word occult reverberated through the halls of a science that was attempting to establish itself on the basis of objective inquiry.

The cause of the force might well be occult, but the calculations performed on the moon dealt with matter so that the cause had to be as real as the matter associated with the cause.

Gravity had to exist as something in reality!

Not as an occult concept!

To defend against the claim that gravity as a property of matter was indeed an occult concept, the Newtonians eliminated slander from rational scientific discourse, a noble undertaking, and then labeled the use of "occult" as a slander of the worst ilk.

They then branded any non-Newtonian thought as occult!

We can see the process of the equilibrical mind functioning perfectly as it attempted to eliminate that in reality for which it could not produce recall reality by creating a recall reality that could become a shared reality, and reducing opposing realities to ridicule.

The quality of our science, our attempt to come up with a consistent picture of physical phenomena, is dependent on the extent to which we defend our conclusions as being evident to the sense of all mankind, as Newtonians are wont to say.

When something is evident to the sense of all mankind, it is not science.

Science must be comprehension rather than mere knowledge.

Statements such as an object will travel in a straight line unless a force acts upon it, the universe is running down, matter can neither be created nor destroyed, these statements are all non-provable.

When your methodology is based on procedures that only disprove facts, hypotheses that cannot be disproven become irrefutable facts.

When we try to prove hypotheses with a process that can only disprove facts, we turn the universe on its ear, seeing things that aren't there and not seeing things that are, simply because we can create a recall reality that is more comfortable than the conflict not comprehending creates.

And, if doom is not imminent, if the organism is not immediately threatened by its failure to comprehend, the conflict does not drive the organism, it provides it with an easy chair.

What we don't know won't hurt us, it will just eliminate us!

Hypotheses are necessary because they are the spatula with which we flip our concepts as we examine them and arrange them so that they will explain physical phenomena.

But laws and properties and principles, axioms and all of the other concepts that go to underpin mathematical equations so that what goes in comes out in an elegant manner, elegance being the touchstone of proof for all of the nonsense that comes from relying on the tools we created to measure physical phenomena to understand physical phenomena, are just methods of stopping the effort at comprehension.

When we create a law, or assign something as a property of something else in order to explain, we are just creating a recall reality that eliminates conflict, and allows us to relax in the face of our failure to comprehend.

The laws are markers for our failure to understand physical reality.

When we say that the moon is traveling in a straight line unless another force acts on it, we are saying we don't have the foggiest idea what is making the moon travel in a straight line in the first place, to attempt to find out would cause us conflicts that would make us feel uncomfortable, so we are going to accept an explanation that makes us comfortable because we don't want to feel uncomfortable.

When we say that energy can neither be created nor destroyed, we are saying that we haven't got the foggiest idea what energy is, and we don't care.

When we say that gravity is a property of matter, we are saying the we haven't got the foggiest idea what gravity is.

To consign physical phenomena to laws and properties is to mark our ignorance.

But we have to use assumptions and conclusions at some point.

How do we get around marking our ignorance?

We don't!

We are always going to be ignorant about something.

The idea is to push the things that we are ignorant about to the very edge of existence, and then, and only then to assign properties to our ignorance.

This is basically what I have done in this book.

The edge of existence is a hypothetical elementary particle, and the properties are merely the manifestations of the matter we live with, the matter the elementary particles have to make up.

IF WE HAVE TO ASSIGN PROPERTIES, IF WE HAVE TO PROCLAIM OUR IGNORANCE, WE MIGHT AS WELL DO IT AT THE MOST ELEMENTAL LEVEL, THE BASIC PARTICLE THAT MUST UNDERLY EXISTENCE.

We will never know whether in fact the elementary particle I hypothesize in this book exists, or whether it evidences the properties that I have induced for it.

It is merely a convenient tool which can be used to explain the operation of physical phenomena the way that we observe and measure it.

When I was in school, it was popular to question existence, to ask for proof that life was other than but a dream. If the questioner could be induced to bend over with his head several inches from the nearest wall, proof was not long in coming.

There actually are some things that are evident to the sense of all mankind.

Existence is one of them.

We exist, we are made up of the same matter that makes up the universe, but we are made up in such a way that we can comprehend not only the matter that makes up the universe, but also the matter that makes up ourselves.

The thing that distinguishes us from other matter is that we can comprehend. To one degree or another, we can look out at the stars, identify the sun as a star, and position ourselves with respect to the matter that populates the universe.

COMBUSTION DRIVES THE FORMATION OF CHEMICAL AND EQUILIBRICAL MATTER. WITHOUT COMBUSTION, THERE WOULD BE NO DISBALANCE IN THE UNIVERSE.

If comprehension is what distinguishes us from other matter, then we must have evolved to comprehend, and as matter is the only thing there is in non-existence, then we must have evolved to comprehend matter.

Which means that we are matter that has evolved to comprehend ourselves.

As we have seen, given an elementary particle with at-rest motion and affinity propensity, this evolution is fairly straight forward.

Elementary particles form with opposing properties,

at-rest motion and affinity propensity.

These opposing properties seek equilibrium and move the elementary particles into positions that form them into physical matter.

The physical matter ignites and combusts.

Combustion is the great disbalancer, throwing matter into perpetual disequilibrium.

Matter constantly attempts to regain equilibrium, breaking down into a variety of physical matter which creates equilibrium on one scale, but removes it on another.

As the matter undergoes combustion, it moves with respect to each other, and the movement creates disequilibrium in the newly formed matter.

The disequilibrium in turn creates movement to equilibrium, the new movement resulting in the creation of chemical matter, matter that can obtain equilibrium in an environment that is being continually disbalanced by the movement of the physical matter in the environment.

It is a process of matter moving into equilibrium only to be disbalanced and forced to regain equilibrium under changed circumstances.

Physical matter in the form of planets and stars move to regain the equilibrium lost to combustion.

The movement creates disequilibrium in the physical matter that makes up the planets, and chemical matter forms on their surface to regain the equilibrium lost to the movement.

As the chemical matter forms to provide equilibrium to the moving physical matter, it finds itself moving independently of the physical matter it formed to provide equilibrium to, addressing the disequilibriums in a fluid environment.

Because it had to move within that environment, the chemical matter had to evolve a way to sense that environment, doing so by creating equilibrical matter, elementary particles in equilibrium with each other without being immobilized into physical matter.

IF EVOLUTION IS A PROCESS OF DISEQUILIBRIUM SEEKING EQUILIBRIUM, THEN OBTAINING BALANCE BY ACCEPTING SHARED REALITIES THAT HAVE NO BASIS IN REALITY IS COUNTER-EVOLUTIONARY. IT DEFEATS THE WHOLE PURPOSE OF EXISTENCE, THE GOAL OF COMPREHENSION.

The resulting equilibrical matter could sense the environment by using the emissions from the combustion that were disbalancing the physical matter to start with.

Which raises the central question about comprehension: Is comprehension merely a byproduct of the ceaseless effort of the elementary particles to balance their opposing properties of at-rest motion and affinity propensity, the result of the long process of attempting to achieve an unobtainable equilibrium, or did the elementary particle itself evolve so that it had the right mix of at-rest motion and affinity propensity so that it could develop awareness, the ability to comprehend itself?

It is not an idyll question.

If awareness is merely a byproduct of the interaction of elementary particles to regain equilibrium in response to environmental conditions that lead to disequilibrium, then there really isn't a purpose to awareness.

Nothingness, the non-existence of matter, is perfectly balanced to start with, if something that has no existence can be refereed to as balanced.

In nothingness, total non-existence, there is no movement, no space defined by matter, no thought, simply nothing.

Our observation tells us that nature abhors a vacuum, so it would be easy to say that matter abhors non-existence and therefore came into existence.

However, it is easier to believe that because non-existence simply doesn't exist, it is the absence of anything, that matter would just come into existence.

After all, when a condition in which there is stationary matter against which movement can be measured, then the interval really defines the time available for matter to occur.

In nothingness, that time is infinite.

In an infinity of possibilities, it is probable that the improbable will occur.

But, in the end, it is like looking at the sun and saying it is giving off energy which is dissipating in the universe, or saying that the sun is adding energy to the universe, energy which will dissipate, but which will form once again into matter which will repeat the cycle endlessly.

By assuming that the elementary particle evolved the properties that produced awareness, we assume that there is a purpose in comprehension.

It is possible to conceive of the elementary particle forming with a little too much affinity propensity, in which case the structure it created would not be limited by ignition, but would continue to grow in size as long as elementary particles came into existence with too much affinity propensity.

It is also possible to conceive of the elementary particle forming with too little affinity propensity, it shooting off into non-existence immediately, never forming into even a miniature unit of matter.

After all, there was an infinity of time in which to make an infinity of mistakes, and an infinity of space to make them in.

It is also easy to conceive of an elementary particle that has just about got it right, but not just right.

It is conceivable that such a particle could build a universe, that it could form into stars, the stars could ignite, the ignition produce combustion, the combustion produce emission fields which reduced matter and caused it to move so that it would produce the systems of awareness, the chemical matter that balanced the movement, the equilibrical matter that balanced the chemical matter as it moved within the context of the moving matter.

It is conceivable that such an elementary particle was able to evolve into complex structures that could become aware of the universe, just like we are aware of the universe.

BECAUSE AN ELEMENTARY PARTICLE HAS FORMED WITH THE RIGHT PROPORTION OF AT-REST MOTION AND AFFINITY PROPENSITY TO FORM THE UNIVERSE WE EXIST IN DOES NOT ALLOW A CONCLUSION THAT IT WAS FORMED WITH THE RIGHT PROPORTIONS TO COMPREHEND ITSELF.

And it is conceivable that because of a defect in the working of the equilibrical structure with respect to the chemical organism, the elementary particle that has evolved is prevented from actually comprehending itself, prevented from understanding how elementary particles work to form the universe that exists.

In such a case, the mix of at-rest motion and affinity propensity that made up that particle would be the wrong mix, the equilibrical structures developed by its interaction would never result in the comprehension necessary for the existence of the elementary particle to begin with, and that universe would drop by the wayside to make room for another one made out of the same elementary particle with a slightly different mixture of at-rest motion and affinity propensity so that when it finally reached the point at which an equilibrical structure developed to comprehend physical phenomena, it wouldn't be satisfied with half assed comprehension.

Having an understanding how comprehension operates makes it clear that it is easy to accept physical reality, and alter recall reality to the extent necessary to come up with a consistent picture of physical phenomena where there is an actual desire for comprehension.

I don't think that we are a product of a failed evolution of the elementary particle.

We've just failed to realize that our purpose is to comprehend, and that the purpose of comprehension is to comprehend ourselves.

For one thing, the equilibrium of the equilibrical structure, our mind, is based on a non-physical structure that is unaffected by the basic cycle of the universe, formation, ignition, combustion, emission, and dissipation because it is not subject to the attractive mechanism embedded in the expanding fields of electromagnetic emissions.

It is the attractive mechanism that drives the cycle, sooner or later pulling all matter into the combustion process so that it can participate in the cycle over and over again.

To the extent that the equilibrical structure is not subject to the attractive mechanism, it has the endurance necessary to the long process involved in comprehending physical phenomena.

However, there is an even better reason to believe that we are not the product of a failed experiment.

Our closed end mathematical process of analyzing physical reality always results in closure because the equations have to be self sufficient and balance.

Nothing else might be true of closed systems, but closed conclusions always result.

Thus, propulsion systems have to work against gravity and thus would not be efficient over long distances, and if we could find propulsion systems that were good over long distances, the distances are too far to travel, and if we tried we couldn't travel very far anyway because we can't go faster than the speed of light, and switching to the philosophical, we wouldn't find anything if we got there anyway because life is the exception rather than the rule, and come to think of it the universe is the exception rather than the rule, existing in a very narrow band of probability, and on and on and on, spinning endless cobwebs of the impossible.

Without going into the preposterous proclivity of assigning impossible distances for the universe, and impossible dates for the events that created the solar system, the Earth, and the life on it, propulsion systems are possible that not only eliminate the effect of the attractive mechanism, but utilize its infinite acceleration possibilities; Matter can travel faster than the speed of light; It can get to other star systems easily; and when we build the craft to get there, we'll find something there once we arrive, assuming that our organisms can find some way to compensate for the loss of the disequilibrium that gave them rise, the movement of matter resulting from the motion of the planet and the movement of the flows among the chemical matter that makes us up in relation to that motion.

After all, since we exploded the first atomic bomb, our skies have been filled with vehicles operating without regard to the attractive mechanism.

We can't see them because they don't, and can't exist in our recall reality.

It doesn't mean that they aren't real, just that we have no way to comprehend what we see.

Why don't they drop in and say hello?

If you were a space traveler, and you came upon a planet that was inhabited by people who participated in the primitive worship of an invisible force mysteriously gluing them to the ground beneath their feet, and who deified that force as the very essence of reason, would you want to drop in?

Not if they could set an atom bomb off under you, you wouldn't.

APPENDIX

From THE COOLING CONTINUUM: THE RISE AND FALL OF SPECIES ON EARTH, Volume Two of The FBP Copernican Series.

If you believe in ghosts, reincarnation, unicorns, life after death, leprechauns, in fact, in anything, you have more proof, "scientific", or otherwise, for the belief's existence than if you believe that there is something called mass, and that that something causes gravity.

That modern monetary magician, John Maynard Keynes, thought that Newton, the person held responsible for the mass/gravity concept, was the last of the great magicians.

Newton was a magician of sorts. As the leading alchemist of his day, he represented that group of men that spent their adult years trying to turn lead into gold.

As a mathematically gifted alchemist, he was interested in the measurement of phenomena in physical reality.

The inverse nature of the force that caused objects to fall had been identified before Newton came on the scene.

*****

Mass was necessary to prove the inverse square law, but the inverse square law is necessary to determine mass.

Newton didn't create the modern concept of mass. He just came up with the idea that the attractive force was some sort of magical property of the matter whose movement allowed us to indirectly measure it.

When Newton was challenged on this belief, when it was pointed out to him that the attractive force was not even deducible from the existence of matter, let alone associated with it, he replied that for that reason, for the very reason that we don't know what the attractive force is, we have to use an occult reason to explain it.

How is it that an occult concept used to explain a force as obvious as the force that governs our every movement, concocted by a seventeen century alchemist has come to be the central belief of western civilization?

The answer to that may be found in John Maynard Keynes' domain of economics.

In rummaging through history, the same principles apply that are guiding us when we attempt to conceptually order the past of the Earth to find out more about ourselves. The history is gone. We have to reconstruct it as best we can.

Fortunately, the history surrounding making Newton, and thus his thought, "[t]he great ornament of the Country and Light of the World" is not too hard to reconstruct.

From the defeat of the Spanish Armada in the late sixteenth century, scarcely a hundred years before the Newtonian ruminations that have confused the effort to come up with a consistent explanation for physical phenomena, the British ornament of Empire was rising as the light of the world.

The Empire was based on mercantilism. Mercantilism was the simple process of subjecting the people of the world, and the resources they commanded, to control through the control of the medium of exchange.

Put more simply as the golden rule, he who controls the gold, rules.

In Britain's case, empire came with sea power, which came fast after the defeat of the Spanish. The resulting trade in a world quickly grown small with the control of the shipping lanes soon outgrew the supply of gold and silver available to conduct it.

Mercantilism meant that trade must always produce a surplus of the medium of trade, gold and silver, in favor of Britain.

By the end of the seventeenth century, the failure of the gold and silver stock to increase in proportion with the increase in trade threatened the continued growth and prosperity of the Empire.

As a result, it was necessary for the British rulers to come up with an alternative to gold and silver that would facilitate the burgeoning trade without endangering Britain's position of control of the gold and silver itself.

The answer was the Bank of England, formed in sixteen hundred and ninety four. It can safely be stated that the question of whether the moon was falling at the same rate as a rock, or whether it was falling because of some self contained principle, was, in sixteen hundred and ninety four, at best a question of academic interest to those academically inclined, but certainly not a point central to the existence of the Empire.

The way that the Bank of England was operated would soon change that situation because The Bank, as it was to become known, was the first stable national reserve bank. A reserve bank creates credit on the basis of the gold and silver it holds in its vaults. Because it is The Bank, it can create credit even though it doesn't hold the gold and silver to back that credit.

The Bank of England was able to facilitate the growth of trade beyond the limitations of the existing gold and silver supply because it was able to substitute its credit for the gold and silver that would otherwise have been required for the transactions that underlay the trade.

Reserve banks are not only common today, the business of the world could not be conducted without them.

But in the world of sixteen hundred and ninety four, and in many areas of the world today, people preferred to have the gold rather than the promise of the gold.

In addition, there were established dealers in gold that didn't look kindly on a gold substitute, and in whose interest the destruction of the Bank of England was foremost.

The Bank of England had hardly opened its doors before the forces bent on its destruction were busy obtaining legislation that would require the Bank's founder, William Patterson, to underwrite the issuance by the Mint of gold coins containing full weight in exchange for coins which had been clipped.

In addition to requiring Patterson to come up with a significant amount of gold to replace the gold that had been clipped from the currency, the passage of the act would create panic among note holders, whose notes exceeded the gold and silver in The Bank. They would all want their gold and silver at the same time, and the resulting panic would destroy the fledgling bank.

Patterson was faced with the problem of how to ally the fears of depositors who could clearly see The Bank's failure in the new legislation.

How could he change a bad belief, that there was no gold to back up the credit of The Bank (which was true), to a good belief, that the mint could create more gold than would ever be needed (which was false).

Put an alchemist on the payroll!

It didn't matter that it wasn't possible to turn lead into gold. All that mattered was that people believed that you could turn lead into gold. After all, credit itself is a matter of belief, a belief in the existence of the gold to back up the credit. All that was necessary was to counter a bad belief with a good belief.

The events that were leading to doom for the newly formed Bank of England were coming to a head in the first four months of sixteen hundred and ninety six. Newton was offered the post of Master of the Mint in late March, and was moved to London and installed before the end of April just in time for the panic caused by the redemption of the clipped coins to hit in early May.

The run on The Bank was averted, and history tells the rest, with

The Bank of England bankrolling merry old England to Rule Britannia.

However, it didn't do much good to have Newton as Master of the Mint unless the world realized that Newton wasn't just any Master of the Mint. He had to know the laws of creation. He had to be one step beneath the deity that set creation in motion.

Thus, in seventeen hundred and three, Newton was made the President of the Royal Society. William Whiston was dispatched to control the Cambridge front, while Edmond Halley (of comet fame) did the same at Oxford. Halley, along with James Jurin and John Machin were made officials of the Society to control the flow of its business, which is to say, to determine what were valid thoughts and what were invalid thoughts. Francis Haukesbee and J.T. Desaguliers were put in as the verifiers of experiments so that there would be control over what were facts and what were not facts.

With the levers of power in hand, a bogus doctrine of chemistry was created to demonstrate the reality of Newton's occult belief that the attractive force was seated in matter itself, and that by implication, Newton's knowledge put the secrets of the universe, one of which, of course, was the transmutation of base metals, in British control.

Haukesbee and Desaguliers conducted specious experiments on a continuing and extremely public basis that conclusively proved the connection of gravity to matter through such parlor tricks as particle cohesion, capillary action, and the electrical effect exhibited by matter, explaining away the measurable discrepancies of the latter as proofs.

Samuel Clarke stimulated and maintained an international dispute with Leibnitz as to the source of Newton's calculations, and the two John's, Freind and Keill tried to suck Leibnitz into arguing about the bogus alchemical based doctrines that were at the basis of the deification process. (Leibnitz, who had at one point identified the attractive force as something that had to be similar to the light streaming out from the stars, ended up with theories using as analogies vortices derived from watching matter flow in whirlpools, a concept that at least had the benefit of using a current and existing, as opposed to historical and non-existent force to deal with orbiting.)

Henry Pemberton and Colin Maclaurn made sure everything was properly promulgated by churning out a continuous flow of textbooks to make it all seem real (echoes of our present day glossies which graphically depict the worm holes connecting the gravitational vortices around black holes).

Thus did the powers that be construct Newtonianism, the deification of an alchemist as the Master of the Mint, not just an ordinary alchemist, but an alchemist that knew the deepest secrets of nature. There could be no question of the credit of The Bank of England, when England had within the stone walls of its mint a man that could turn base metal into gold.

From AT THE GATE OF THE CITADEL: THE SUBJUGATION OF MODERN SCIENCE, Volume One of The FBP Copernican Series.

The adoption of the British deductive methodology, where non-mathematically disproven concepts are accorded the status of fact, leads to non-understanding.

As science is understanding, the deductive methodology is not scientific.

Because we cannot understand physical phenomena without the use of concepts, we cannot rely on a methodology that does not accord credibility to concepts other than on the basis of their ability to withstand disproof.

Because the young British Empire had to control knowledge, it had to use science as an instrument of state power.

The infallibility of British science was embodied in the promotion of Newton as the person who had divined the basic law of the universe.

If the Empire's science had divined the basic law, then it was the arbiter of all other laws.

Because laws are not subject to proof, being the arbiter of law allowed Britain to dictate what was science and what wasn't science.

This was done through the deductive methodology, the so called scientific method, which allowed the Empire to dictate what was a valid hypothesis, what was a valid method of proving the hypothesis, and what constituted the finding of fact that provided the proof of the hypothesis.

Because, under the British deductive methodology, proven hypothesis became scientific fact, the Empire could dictate what was and was not a fact.

*****

The only thing that Newton had to start with was a rock dropping in accordance with an inverse square law, the distance of the moon from the Earth, its orbital period and the desire to prove that the moon was dropping in accordance with an inverse square law.

He assumed how much matter was in the moon, and he assumed that the moon would travel in a straight line but for the fact that it was dropping like a rock.

The first assumption is beneath reproach. It is beyond determining. The second assumption, that the moon would be travelling in a straight line but for the force that was making it drop like a rock, requires so many more assumptions to keep it valid that it becomes absurd. The assumption that the planets would travel in a straight line but for the force of gravity requires the assumption that there is a certain amount of inertia in the system which requires that the first assumption, that we can determine how much matter exists in the system, be above reproach.

If we get rid of all of the assumptions, then we are left with simple facts to explain: Why do planets orbit around the sun; why do the moons orbit around the planets; why do they all rotate; why is all of this motion generally in same direction; and why does it all occur in the equatorial plane of the sun.

When you get rid of the facts that are actually concepts, and the assumptions that have been accepted as facts, the number of facts left to explain becomes small indeed.

And those facts are the very facts that the British Methodology has forced us to ignore simply because they don't agree with the mass/gravity concept, the idea of gravity as a property of matter.

Sir Francis Bacon preceded Newton and Bacon was definitely a person that favored Ockham's razor, the concept system just described where that which explains the most with the least is the best. This inductive methodology requires that understanding come from verified facts.

Why did the deductive methodology become the dictated methodology?

Because knowledge is power, it was necessary to have knowledge centered at the seat of the rising British Empire.

The deductive methodology is a methodology designed to control knowledge.

On an objective basis, the importance of Newton's measurements are minuscule when compared to Copernicus' re-arrangement of the then current observation. However, by creating a god of Newton, who was after all not much more than a mathematically gifted alchemist, Britain could elevate its claim to knowledge above the efforts of all other nations.

And in doing so, it could define for other nations what knowledge was.

By dictating the content of knowledge, it could funnel the

discoveries of other nations through its nationally formed organizations, validating or suppressing them as the interests of the state dictated, but more importantly, stealing with impunity.

In the United States, our engineers have always used the inductive methodology. Our technology has always preceded the concepts that we used to explain it. The transistor is a perfect example, with its concepts of moving holes being, in the British tradition, incomprehensible.

Historically, the best example of the total theft of theory and invention involved the discovery by Joseph Henry of electrical inductance. Henry had not only discovered inductance, but had put it to practical use in a telegraph by eighteen hundred and thirty.

The British Empire, however, using its position as arbiter of scientific validity derived from Newtonian infallibility, claimed that the discovery was made in eighteen hundred and thirty one by Faraday, another person whose arts, like Newton's, were based in the chemicals.

Today, we politely acknowledge a joint discovery, an absurdity that is in accord with the joint "discovery" of Newton's calculus.

The British Methodology results in non-understanding simply because, being an instrument of state power, it is designed to produce non-understanding.

However, whether it was felt at one time that understanding could be produced nationally while confusion was spread internationally, the result has been universal: Total non-understanding on a universal basis.

MATTERS INHERENT IN MATERIAL BUT COVERED IN FIRST TWO VOLUMES

The cooling planet drives the evolutionary process, with life forming in a much hotter environment, and going through successive waves as the cooling continuum allows new life forms to develope while eliminating existing species as the temperature drops below the capacity to survive.

Field replacement is how matter is broken down. When units of matter are slammed together in accelerators, and then interpreted in accordance with Newton's second law of motion, all forms of matter are possible, and all interpretations are available for the created forms.

The concept model is the subject of Volume One. Basically the process, instilled in the process of factual congruities in Volumes Two and Three, entails deriving concepts for sets of facts, and then comparing those concepts, using overlapping facts in the sets to refine the concepts. Those concepts which explain the most facts with the least number of propositions, exceptions or reservations are the most valid.

The inductances associated with flows of elementary particles making up the construction that we measure as the electromagnetic spectrum are easily measurable as the magnetic fields on the surface of the sun, where light is at its highest concentration.